DOUBLE HELIX HISTORY

Double Helix History examines the interface between genetics and history in order to investigate the plausibility of 'new' knowledge derived from scientific methods and to reflect upon what it might mean for the practice of history.

Since the mapping of the human genome in 2001, there has been an expansion in the use of genetic information for historical investigation. Geneticists are confident that this has changed the way we know the past. This book considers the practicalities and implications of this seemingly new way of understanding the human past using genetics. It provides the first sustained engagement with these so-called 'genomic histories'. The book investigates the ways that genetic awareness and practice is seemingly changing historical practice and conceptualisation. Linking six concepts – 'Public', 'Practice', 'Ethics', 'Politics', 'Self', and 'Imagination – *Double Helix History* outlines the ways that genetic information, being postgenomic, the public life of DNA, and the genetic historical imaginary work on the body, on collective memory, on the historical imagination, on the ethics of historical investigation, on the articulation of history, and on the collection and interpretation of data regarding the 'past'.

This book will appeal to researchers and students alike interested in DNA, genetics, and historiography.

Jerome de Groot is a Professor of Literature and Culture at University of Manchester and he was the Chair of Manchester UNESCO City of Literature consortium from 2016 to 2019. He has authored a number of historical books, including *Remaking History* (Routledge, 2015) and *Consuming History* (Routledge, 2008/16).

DOUBLE HELIX HISTORY

Genetics and the Past

Jerome de Groot

Routledge
Taylor & Francis Group

LONDON AND NEW YORK

First published 2023
by Routledge
4 Park Square, Milton Park, Abingdon, Oxon OX14 4RN

and by Routledge
605 Third Avenue, New York, NY 10158

Routledge is an imprint of the Taylor & Francis Group, an informa business

British Library Cataloguing-in-Publication Data
A catalogue record for this book is available from the British Library

Library of Congress Cataloging-in-Publication Data
A catalog record has been requested for this book

ISBN: 978-0-367-51236-1 (hbk)
ISBN: 978-0-367-51235-4 (pbk)
ISBN: 978-1-003-05297-5 (ebk)

DOI: 10.4324/9781003052975

Typeset in Bembo
by MPS Limited, Dehradun

For Gen, who loves to make the rules

CONTENTS

FIGURES

ACKNOWLEDGEMENTS

The following have helped me in various ways through correcting my ignorance, supporting my work in various thoughtful ways, or opening my eyes to new ideas. I am grateful to their generosity over the past few years: Matthew Cobb, Dan Davis, Phil East, Peter Wade, Jackie Stacey, Rich Willock, Gagun China, Alison Light, Ben Harker, Scott Midson, Clare Barker, Zaffar Kunial, Laura King, Naz Kahn, Chris Wallace, Paul Knevel, Dave Govier, David Langrish, Nicola Phillips, Carolina Jonsson Malm, Sara Trevisan, Ewa Jurczyk-Romanowska, Brian Gallagher, Stefan Krankenhagen, Andreas Fickers, Kees Ribbens, Rachel Pope, Hayley Dunn, Robert Witcher, Berber Bevernage, Chris Lorenz, Hans Ruin, John Marsden, Manchester and Lancashire Family History Society, John DiMoia, Leslie Turner, Erin Battat, Sarah Abel, Hannes Schroeder, Nathan TeBokkel, Marc Scully, Henriette Roued-Cunliffe, Marcelo Abreu, Andre Freixo, Indira Chowdhury, Robert Witcher, Josie Gill, Malcolm Allbrook, Ruth Connolly, Kerry Farmer, Nathan Scudder, Paul Pickering, Helen Curry, Martin Willis, David Kirby, Carsten Timmerman, Michihiro Okamoto, Scott McCracken, W. Patrick McCray, Jonathan Pegg, Serge Noiret, Eun Kyung Min, Sophie de Carné, Kier Waddington, Joseph Yracheta, Krystal Tsosie, Turi King, Catherine Nash, Adam Rutherford, Muriel Bailly, Jonathan Pegg, Sarah Williams, Bryony Partridge, Kristen Hyde, Abigail Saffer, Darryl Leroux, David Dean, Georgina Wells, Reece Williams, Angela Davis, James Sumner, Emma Britain, George Burghel, Ian Scott, Dee-Ann Johnson, Nicholas Weise, Stefan Nyzell, Maya Sharma, and Jon Dickson.

Thanks to those who generously read sections of the book before publication: Thomas Booth, Iain Mathieson, Christopher Vardy, Kalle Pihlainen, Marnie Hughes-Warrington, Jay Clayton, Lara Choksey, Eloise Moss, Debbie Kennett, Cathy Day, and Thomas Cauvin.

Matthew Stallard was my RA at times and did loads of great work for various projects, taming my wilder ideas, and figuring out how to make things happen. Thanks so much Matt! Thanks to Eve Setch, Zoe Thomson, and Louise Ingham at Routledge for supporting the project and ensuring publication was smooth.

In Australia, many people made me welome and generously gave me help. Particular thanks to Tanya Evans and Ashley Barnwell for fun times and difficult questions. I should thank Penny Wolf at the Genealogy Society of Victoria, Clare McGuiness at HAGSOC in Canberra, Danielle Lautrec at the Society of Australian Genealogists, and many of the members of those organisations for participating in conversations and workshops. I spent fruitful time at LaTrobe University with Kelly Gardiner and Catherine Padmore. Anna Clark, Hsu-Ming Teo, Marnie Hughes-Warrington, Stephen Foster, Catherine Frieman, Cathy Day, and Paul Kiem were welcoming and challenging (in the right way). I spent time in Holland and thanks for a great welcome from John Boeren, the Nederlandse Genealogische Vereniging, and Katy Barbier-Greenland. In Tokyo, Rieko Suzuki, Alex Watson, and Tetsu Yokamoto made me very welcome.

Thanks to International Federation for Public History, International Network for the Theory of History, the Chemical Heritage Foundation, Ancestry.com, Science+Industry Museum Manchester, Arvon Centre Lumb Bank, Manchester Central Library, Manchester Museum, the Wellcome Trust, the British Society for Literature and Science, and Ahmed Iqbal Ullah Race Relations Centre.

Much of the groundwork for this book was made possible by a Leadership Fellowship from the AHRC. I spent some time as a Research Fellow at the Humanities Research Centre at the Australian National University, for which I am very grateful (thanks to Will Christie and Penny Brew there). I spent time in Melbourne thanks to the University of Manchester's Manchester-Melbourne fund. I spent time in Tokyo as a Visiting Fellow at Waseda University's Institute for Advanced Study, thanks to Makiko Azami there.

I've talked about this subject in Belfast, Melbourne, Sydney, Canberra, Stoke, Tokyo, Krakow, Ouro Preto, Leeds, London, Berlin, Salzburg, Malmö, Oxford, Bristol, Newcastle, Wroclaw, Amsterdam, Solihull, Bengaluru, Paris, Seoul, and Manchester; thanks to everyone who asked questions and challenged my thinking.

And Sharon, you're the best thing that ever happened.

PERMISSIONS

Short sections of Chapters 2 and 6 appear in my essay 'The DNA Archive' in Zoltán Boldizsár Simon and Lars Deile (eds.), *Historical Understanding: Past, Present and Future* (London: Bloomsbury, 2022), with permission.

Part of Chapter 5 appears in my article 'History and pastness in the post-genomic poetic imaginary', *Medical Humanities* 47 (2021), 177-84, with permission.

INTRODUCTION

In June 2020 a research group published work that investigated the genetic make-up of the Dead Sea Scrolls.[1] The fragmented nature of these documents, consisting of *c*.25000 pieces that are parts of around 1000 manuscripts, means that identifying connections between each piece can make a significant difference. Created between the third-century BCE and the first-century CE, the Scrolls were made from various animal skins. This meant that DNA samples could be harvested from them and analysed. The data yielded enabled scholars to make assertions about the sources of the material use to make the manuscripts and hence how and when they were composed. The authors of the new research argued that their 'paleogenomic approach can shed new light on a range of issues pertaining to DSS history, including the compositional history of the biblical book of Jeremiah, the dissemination of the *Songs*, and the provenience of the scrolls at large', suggesting a contribution to biblical scholarship, historical knowledge, and a reconfigured understanding of the entire collection of ancient manuscripts (p. 1220). The genetic information produced enabled a different 'reading' of the scrolls, expanded knowledge about them, raised hitherto unforeseen issues, and generated important new contexts for interpretation. Researchers were given an entirely original way into understanding and reading, a fresh means of interrogating the materiality of the text. This innovative evidence resituated these important and well-studied manuscripts; 'new light' allowing new understanding. The Dead Sea Scrolls have been worked on by hundreds of scholars over decades, but this approach to the physical aspects of the texts opened up a different way of thinking about them: 'We can suddenly see things that were not visible using more traditional historical, archaeological or literary sources'.[2] Genetic analysis – on this reading – provides a new way of looking at, and an innovative mode for investigating, the past. DNA work here gives a 'fourth' way of engaging with the past, challenging and augmenting 'traditional' modes of investigation. The novelty of this way of seeing is reflected in the 'new light' language

DOI: 10.4324/9781003052975-1

and title of the original research article itself ('Illuminating Genetic Mysteries') and press accounts ('Ancient DNA Yields New Clues'). The language used here is of revelation and detective work, of unearthing and discovery, and of modernity and novelty. The Dead Sea Scrolls, then, might be reconsidered, reconfigured, and reread – all enabled through the intercession of 'new' genetic techniques. Historical event, process, and practice is transformed. This example, and hundreds of others like it, invites questions about the 'new' genetic history that has being enabled over the past two decades.

This book emerged from a conviction that the explosion in genetic knowledge and awareness of the last two decades allows us to ask new questions about history and the ways in which communities comprehend and engage with their past. What are the implications of expanded genetic datasets for the practice of history? How can genetic data change experience of the past, and our way of conceiving what that past is? How does genetic knowledge challenge normative versions of what historical information might be, or how it might be presented? 'DNA analysis is re-creating how we know the past and even how we now define the social world' argue Alondra Nelson, Keith Wailoo, and Catherine Lee. In what follows I seek to understand and to a certain extent critique this 'recreation' in relation to historical understanding, practice, and imagination. Throughout I argue that the impact of genetic science affords us an opportunity to reflect upon and rethink historical practice and historiography. Somewhat fittingly, given the fluid yet structured nature of the double-helix itself – at times throughout the book conceived of as real and unreal, physical and imagined, powerful and fleeting – my account twists strands of thinking. The concept of 'double-helix history' allows me to describe a multiple model. Revolving around a central axis and always fluctuating, the book elaborates ways of considering genetics in relation to history: as a means of producing and creating new data about the past; and as a way of rethinking modes of engagement with the past. The book provides an outline of the implications for historical knowledge and awareness of the explosion in genetic knowledge over the past two decades. Concomitantly 'double helix history' is a set of suggestions for how genetic information might challenge and change the practice of history. Each strand is interwoven with one another and linked in a variety of ways; each is delicate but strong. Of course, this DNA metaphor is imperfect; but the strength of the double-helix as a conceptual and imaginative template for historical thinking is highly important, and enduring, as I show throughout.

DNA, in my argument, is both a tool and a mode of knowledge. 'Double-Helix History' is both an account of the way that genetics has been used to expand historical knowledge and a reflection upon how that new knowledge might shift our understanding of what history is and how it is produced. The book is not meant to survey the meaning of genetics in culture or to generalise about the representation of DNA – many others have done this – but to reflect upon the ways in which genetics might shift our engagement with and experience of the past in the present.[3] Throughout the book I consider the practicalities and implications of a seemingly 'new' way of understanding the human past using genetics. The book contributes to

investigation of what Jackie Stacey terms the 'genetic imaginary' and seeks to elaborate what Alondra Nelson calls the 'social life of DNA'.[4] Whilst this imaginary and
social life in the contemporary has been well studied, the task of this book is to place
such thinking within an understanding of what Nadia Abu El-Haj names 'the genetic historical imagination', something which has been considered much less
often.[5] To date there is little engagement on the part of historians with the practice
and implications of contemporary genetic science.[6] We lack a working methodology for thinking about how DNA investigation impacts upon historical
knowledge, practice, and imagination.[7] DNA and identity has been investigated by
various important scholars but rarely considered as a phenomenon that relates to 'the
past' and to history.[8] Few writers have considered what genomics could mean for the
practice of history or considered how it impacts upon public understanding of
the past. Similarly, few scholars have considered whether geneticists should be thought
of as historians, or how their work alters our understanding of the past. In this book
I elaborate multiple ways that genetic information appears to have changed, augmented, and challenged historical knowledge and practice. The book outlines a set
of approaches for comprehending and understanding the ways that we think about
the past. Consideration of different modes of 'reading' the past reconfigures our
understanding of what that past is, how it is constructed, and what it might mean.

As can be seen in the case of the Dead Sea Scrolls, the claims made on behalf
of genetics for history are wide ranging. Genetic investigation seems to unlock
secrets, enable new knowledge, and reveal information thought lost or un-
recoverable. It can enable a user to experience a bodily connection to those long
dead, through genetic genealogy. It can fundamentally change the way that the
past is considered, by presenting new information in the form of ancient DNA
data. It can enable us to imagine a new type of humanness, through challenging
and changing our understanding of the biological body. Many commentators are
confident that new information has changed the way we conceptualise and articulate history. 'There is now another way to read our pasts' argues the science
writer Adam Rutherford, insisting that DNA has been 'transformed into a historical source, a text to pore over'.[9] Steve Jones argues that that 'DNA opens a
new window into history'.[10] For its advocates, DNA-based work on the materials
of the past is a 'transformative technology [...] providing information that is
comparable in power to archaeology and linguistics' (my emphasis).[11] As will be seen,
it is not always entirely clear exactly what and how, if anything, information
derived from genetic investigation adds to historical knowledge. Yet the rapid
expansion of DNA research and concomitant increase of historical data raises
significant disciplinary and epistemological questions. The surge of work using
DNA, and the 'historical' information it has generated, needs to be considered in
depth and examined thoroughly. On the one hand, historians need to work out a
mode of engagement with this work that allows critique and interrogation of the
basic concepts and language that are being used; on the other, they need to understand the field and develop tools for collaborative working. The implications of
the use of genetics as a type of history might be significant. Genetics-led

investigation of the past puts into flux concepts long foundational to 'history': archive; evidence; narrative; ontology; process; and nation.[12] It may have the effect of obliging historians to recognise the unseen normative assumptions inherent in the way that they approach the past.[13] DNA investigation generates new information, and this needs to be carefully considered; it seeks to present a mode of engaging with the past, which similarly requires analysis. What genetic investigation means about *how* we know the past might be revolutionary. Certainly, *what* we know about the past has developed, and concomitantly the *way* we know that past might be seen to change.

Becoming Postgenomic

The last two decades have been significant and unprecedented for genetic knowledge, both in terms of scientific progress and wider cultural understanding. Work the idea of the gene and on genetic understanding had developed steadily throughout the twentieth century.[14] Francis Crick and James Watson published their description of the molecular structure of Deoxyribonucleic Acid (DNA) which 'has novel features which are of considerable biological interest' in 1953.[15] Crick, Watson, and Maurice Wilkins were awarded the Nobel Prize in 1962 for this work, and increasingly through the 1960s structure of DNA was confirmed and understood.[16] The work of the 1970s focused on sequencing and cloning DNA – what Mukherjee calls 'the "reading" and "writing" of genes'.[17] Frederick Sanger et al outlined a way of 'determining nucleotide sequences' in 1977 which enabled the development of wider technologies to comprehend and understand DNA as it was structured.[18] In 1980 the Nobel Prize for Chemistry was shared between Walter Gilbert/Frederick Sanger (work on base sequences of DNA) and Paul Berg (Recombinant DNA and cloning).[19] Increasingly the idea of the 'gene' and genetics became conceptualised and imaginatively recognised in the public imagination.[20] Lara Choksey argues that 'the genome became a prototype for a new kind of subject'.[21]

This increase in knowledge acquisition around DNA was accelerated in the West during the 1980s. There was increasing institutional and governmental interest in genetic research. In 1985 discussions began about the possibility of sequencing the entire human genome.[22] The Human Genome Project (HGP) would involve determining the approximately 3 billion base pairs that make up the complete DNA of the species, identifying in the smallest detail the makeup of genes.[23] From 1990 the project was given US Congressional approval and funding, with a completion date set for 2005. The National Centre for Human Genome Research (NCHGR) was established with James Watson as director (it became the National Human Genome Research Institute in 1997). A consortium of research institutions around the world contributed to the work, which developed and evolved as the project moved through the 1990s.[24] In 2003 the NHGRI announced that the initial work had been successful.[25] The publication of the report on the project's outcomes and vision for the future was published in *Nature*

50 years after Francis Crick and Watson's original paper on the structure of DNA.[26] The announcement of the new 'epoch' was a self-referentially 'historic' moment, a performance of combined commemoration and futurity. A 'Joint Proclamation' by heads of government from six countries including the USA, UK, and China hailed the 'landmark achievement' in 'decoding all the chapters of the instruction book of human life'.[27] Despite this bombast it was not until 2021 that the 'full' genome was sequenced, but the initial work gave a good profile of the majority.[28] The 2003 outline of the human genome was just the beginning in terms of biomedical advance and development.[29] Multiple new fields and avenues for research developed in areas as diverse as biomedicine, molecular biology, law enforcement, archaeology, evolutionary biology, anthropology, pharmacology, and bioethics.[30] From 2005-7 onwards so-called 'next-generation' sequencing enabled cheaper and faster work to be undertaken on genetic material, leading to fast developments in medicine.[31]

Watson claimed that the decision to pursue the Human Genome project was 'Similar to the 1961 decision made by President John F. Kennedy to send a man to the moon' and that 'the United States has committed itself to a highly visible and important goal'.[32] Watson's rhetoric suggests that, on the one hand, it is a huge, noble, and necessary work of collaboration, and on the other, that it will enable us to 'understand' and 'map' the human. The notion that advanced genetic knowledge might represent a scientific 'moonshot' for the presumed good of all humanity illustrates a (Western) paternalism; as a comparison it also participates in the aspirational, future-led rhetoric increasingly characteristic of statements on genomics, in the face of fracturing faith in such metanarratives.[33] Throughout the project's lifespan it was regularly cited as an impressive and important 'human' achievement. In 1995 the US Department of Energy's newsletter *Human Genome News* noted 'A remarkable spirit of cooperation has facilitated human genome research around the world'.[34] Key was a sense of collaboration and a newly 'big' biological science.[35] Central, too, was the idea that knowing the genome would lead to a 'new' way of understanding the human. Matt Ridley, in his 1999 'autobiography' of the genome, published just before the HGP had shared its findings, wrote that 'In just a few years we will have moved from knowing almost nothing about our genes to knowing everything [...] until now human genes were an almost complete mystery. We will be the first generation to penetrate that mystery'.[36] This latter phrasing echoes scientific statements of revelation, enquiry, and violence from the latter part of the eighteenth-century onwards, and such usage locates the discussion of the HGP within a much longer history of scientific positivism.[37] Concomitantly much public and press discussion of the HGP's purpose and impact was focused around locational and organisational metaphors of maps, exploration, surveys, blueprints, indexes, libraries, and archives.[38] 'Mapping' was the key word, used precisely by the scientists to describe certain processes and then repeated by politicians and journalists alike.[39] Such work had implications for the ways in which culture and society conceived of scientific progress, and in particular suggested a means of 'understanding' life itself.[40] The HGP enabled humans to be 'mapped', comprehended, translated, known, and placed.[41] These motifs of organisation were still to the fore when US President Bill Clinton announced the completion of the major work in June 2000 as presenting 'the genetic blueprint for human beings'.[42] Similarly the NHGRI website presents the project in navigational, exploratory terms:

> The Human Genome Project (HGP) was one of the great feats of exploration in history. Rather than an outward exploration of the planet or the cosmos, the HGP was an inward voyage of discovery led by an international team of researchers looking to sequence and map all of the genes – together known as the genome – of members of our species, *Homo sapiens*. Beginning on October 1, 1990 and completed in April 2003, the HGP gave us the ability, for the first time, to read nature's complete genetic blueprint for building a human being.[43]

The NHGRI description claims that the HGP is itself an historical moment, unique in its scope. The idea of the 'blueprint' is repeated. The Human Genome Project has created a new understanding of the human, allowing 'for the first time' an understanding of the fundamentals of life itself. There are clear colonial echoes in the heroic phrases 'great feats of exploration' and 'outward exploration of the planet', likening the HGP seemingly to nineteenth-century European expansion ('voyage of discovery' similarly has such undertones).[44] That the new human reference archive created by the HGP was predominantly based on European samples has been described as a 'demonstrably supremacist gesture'.[45] Repeated metaphors of discovery, revelation, directionality, linearity, organisation, and rationality associated with the genome and therefore with DNA more generally were predicated upon white, Eurocentric assumptions of progress and civilisation.

One of the key cultural legacies of the HGP is a sense that the human is knowable and comprehendible, compassable and navigable.[46] This seemingly new understanding of the human led to challenges in understanding bioethics, biopolitics, and biopower.[47] Anxieties around genetic engineering, of understanding and reworking the human body from its foundational aspects, of cloning, of the patenting of genes, and the development of biomedical knowledge became standard in the cultural imagination.[48] These anxieties were related to the HGP and the fast development of genetic knowledge. Since the late 1990s the surge in interest in, and popular understanding of, genetics has been enormous.[49] Jay Clayton suggests that in 1999 the *New York Times* 'averaged forty articles a month on the new genetics', and in wider culture genetics and DNA became widely referenced.[50] Niklas Rose articulated the ways in which genetic knowledge was increasingly challenging normativity: 'The laboratory has become a kind of factory for the creation of new forms of molecular life. And in doing so, it is fabricating a new way of understanding life itself'.[51] Other scholars agreed that these shifts were challenging standard political and cultural philosophies, with Paul Rabinow suggesting that 'With the coming of biotechnology, not only new subjectivities and new apparatuses but also new life-forms cross divides 19th-century social thinkers convinced themselves were unbridgeable'.[52] Development in genetic understanding has led to shifts in subjectivity and fundamental ways in which the species is comprehended, with Paul Gilroy suggesting that a focus on the molecular might take focus away from more damaging discourses: 'At the smaller than microscopic scales that open up the human body for scrutiny today 'race' becomes less meaningful, compelling, or salient to the basic task of healing and protecting ourselves'.[53] Many worried that the 'postgenomic' era meant the commodification of genetic material.[54] As Jenny Reardon notes, 'DNA sequencing became a multi-billion-dollar business'.[55] There was a clear sense amongst public and scholars that the human would never be the same again.

Part of what this book outlines is the way that what Reardon calls the 'post-genomic condition' is articulated and interrogated in a number of areas relating to our understanding of the past: public history, imaginative writing and film, commercial development of sequencing, archaeology, and scientific technique. The term 'postgenomic' works in several ways. Principally it is a marker of chronology, a recognition that from the late 1990s onwards our heightened knowledge of our genetic make-up became imbricated within popular culture and historical imagination.[56] This is due to what El-Haj calls 'the epistemological authority of genomics in the contemporary world' (p. 5). Being in Reardon's 'Postgenomic' condition or what Lara Choskey calls the 'age of the genome' or what Siddharta Mukherjee calls 'Post-Genome' – there are multiple ways of articulating it, including 'post-genomic era' and post-genomic age' – is to participate in this social and cultural understanding of the human through a genetic lens.[57] 'Postgenomic' implies a state in which it is impossible to go back 'before' the genome was sequenced, a state that is not necessarily welcome but as it cannot be changed needs to be explored and understood. The sense that the world has been remade or newly made via the intercession of genetic science is deeply felt. The implications of being 'postgenomic' is that everything has changed. If the Human Genome Project was an 'historic' act insofar as it was huge and unprecedented, it was also something of an 'historic' action inasmuch as it created a pivot around which things changed in the world. The idea of a particular moment when society became 'post-' anything is evidently a problematic way of thinking about contemporary life. Dividing into 'pre-' and 'post-' invites the imposing of arbitrary periodisation. The general point, though, is that genetic knowledge exploded in the late 1990s and claimed to reveal increasingly new ways to understand the human. The implications for this are still being debated and discussed, particularly as new developments such as the 'biomedical novelty' epigenetics continue to evolve our understanding.[58]

Knowing Differently

What is the effect of this 'new' status, if anything, on subjectivity, imagination, sensibility, and identity? What does being 'postgenomic' mean to an understanding of, engagement with, and articulation of the past? How does it (does it?)

change practice? This book seeks to understand the ways in which living in a postgenomic world has changed the way that we think about the past. This is not always straightforward, and oftentimes responses to new genetic history are contorted, confused, resistant, and strange. This diversity of reaction is important, and indeed illustrative of the complexity of DNA in the public historical imagination. Michael Symons Roberts articulated the way that DNA seemingly resists binary definition in his 2003 poem 'Mapping the Human Genome': 'Geneticist as driver, down the gene/ codes in, let's say, a topless coupe/ and you keep expecting bends'.[59] Rather than this 'it's dead/ straight, highway as runway' (ll. 5-6). For Symons Roberts the point of DNA is that it is multiply experienced – as helix and as straight line, as concept and bodily reality at the same time, its defining linearity also endless: helix unravelled as vista,/ as vanishing point. Furthermore, DNA defies our attempts to impose meaning and to 'read' and 'see' it. The seemingly contradictory phrase of 'vista/ as vanishing point' suggests that the new maps are for places unseen and unforeseen, on constantly shifting scales. This new flexibility might be astonishing and possibly transgressive; it might also be limiting. The 'vanishing point' is both something that not only establishes order but also invokes endlessness, shifting and moving from compositional to unending in Symons Roberts's thought. It is also a foundational aspect of the art of perspective, which itself was part of the development of a rational mode of looking and ordering in the West, from the fifteenth century onwards.[60] Symons Roberts is one of many authors and poets (discussed in Chapter 6) who took issue with the seeming imposition of meaning associated with the HGP, instead offering ways of thinking about the postgenomic space as something in flux, creative, and possible. I make the case in Chapter 6 that poetry, with its constant elegant resonance between form and meaning, is particularly alive to the flux of the double helix. The idea of the postgenomic need not be governed entirely by metaphors of maps, directions, linearity, and order. Indeed, much of what this book traces is the development of a resistance, or at least a spirited rereading, of the idea of genomics as definitional and fixed. This book considers how this challenge to scientifically-derived models of identity is expressed in considerations of the past, in particular with relation to race, ethnicity, and culture. Particularly acute in discussions of 'postgenomic' identity are suggestions that contemporary genetic science seeks to impose particular frameworks of knowledge that contribute to discourses of control.[61] Discussing James Watson's racist comments the critic Sylvia Wynter argues that he represents 'the enacting of a uniquely secular liberal monohumanist conception of the human'.[62] Wynter's urgent call is for '*challenges* to the single biocentric model of liberal monohumanist Man' (pp. 20-1, p. 23), seeking to disavow the structures of thought that genetic science has attempted to impose. Being postgenomic, then, *might* be to query the seemingly irrefutable truths that are being shared, and in particular in constructing alternative spaces for understanding. Consideration of history and the past on this reading allows writers, film makers, indigenous peoples, and historians to challenge normative biocentrism and enables a critique of the ways that the past is constructed in the present.

Concomitantly the state of being 'postgenomic' might provide a challenge to normativity, a radical overhaul of the ways in which we know, a new epistemology.[63] If the human is known in new ways, the ways in which they engage with the past subtly shift too. The explosion of techniques for investigating humanness after the work of the HGP in the 1990s have led to epistemological and ontological change. Niklas Rose argues that these profound shifts in our knowledge of the species were effecting a change in subjectivity, led by biomedical innovation: 'It is not philosophy but the life sciences which are leading an epistemic change in our relationship to the human'.[64] This challenge transforms our understanding of the past, too, in particular by introducing a newly bodily evidence. David Reich, leading scholar of Ancient DNA, argues that the newly developed techniques might challenge our own understanding of subjectivity: 'It can reveal things about the past that are completely unexpected, that are *not dreamed of in our philosophy*' [my emphasis].[65] Reich's use of Hamlet's famous words signals a desire to re-render the human, to argue that the knowledge that is enabled through scientific technique might allow us to know ourselves differently. Hamlet is taken here to be the advocate of a new understanding, a disclosure that will change our understanding of the past. However, the link is awkwardly made. In the play Hamlet is speaking to Horatio about the Ghost that has been seen and arguing that their shared 'modern' learning might not prepare them for the fantastic or the unknown. It is a moment of mystery, rather than the revelation of new knowledge. The revenant that is the Ghost, the moment of interruption of the modern world by the old, brings a new understanding by reminding the now that the then exists. As Jacques Derrida wrote of the Ghost in *Spectres of Marx*, 'The Thing meanwhile looks at us and sees us not see it even when it is there. A spectral asymmetry interrupts here all specularity. It de-synchronises, it recalls us to anachrony'.[66] Furthermore, there is a famous textual confusion about whether Hamlet says 'your' or 'our'; it is unclear whether he shares Horatio's confusion or seeks to place himself in the vanguard of the esoteric.[67] Reich's reference to Hamlet articulates something particular about the challenge that new genetic knowledge makes to received understanding, particularly about 'the past'. As will be seen throughout this book, genetic understanding seems to constantly reshape epistemology and reorder the ways the past is manifest in the present. Like Hamlet, we are reconfigured by the intervention of something bodily from the past which appears in the present, something that been there all the time yet not seen. Lying dormant in our bodies, DNA seems to become something that can translate the past into the present, highlighting the hitherto unseen historical evidence within our cells.

In his 2000 thriller *Jar City* the Icelandic novelist Arnaldur Indriðason explores the cultural impact of generating and archiving genetic information about human populations, and, indeed, what it means to be 'postgenomic'.[68] The plot revolves around genealogical and health information stored in Reykjavik's Genetic Research Centre. The novel considers the problems associated with genetic science: '"This is all so new for us. People don't understand exactly what can be done with all the information that's been collected. What it contains and what you can read into it.'" (p. 328).[69] New genetic science here acts to reveal unseen and unwished for relationships and connections. What is terrifying is the amount of new information that has been collected, and the possibilities of it being read in hitherto unknown ways. In particular, the data gathered and stored has the potential to change *what* is known about the past, and *how* that past is known. Furthermore, the dread horror of the past might emerge from this data. *Jar City* is obsessed with the past emerging into the present. During a murder investigation in Reykjavik detectives discover a body buried beneath a floor, a man thought to have disappeared years before. A set of photographs reveals indistinct ghostly presences in the background, a woman who might have been the victim of a rape. A girl's body is exhumed and discovered to be missing a brain, removed during a postmortem analysis. In the novel, genetic information becomes a means for violence to erupt in the present, a bridge between then and now but also a means of reconfiguring then in the now, changing the past in the present. Crime fiction often dramatises the horrors of the past exploding with new meaning in the present. *Jar City* mobilises contemporary concern with genetic testing as part of this wider cultural anxiety about the action of the past in the present. Indriðason visits the sins of the father on the father, as the rapist is killed by a son who discovers their relationship because they share a rare genetic mutation. Crime fiction also articulates social fears and outlines the impact of particular new technologies on common understanding. The novel ends with a dire warning about who controls access to the genetic data of the past: 'Erlendur stood up. "And you keep all these secrets. Old family secrets. Tragedies, sorrows and death, all carefully classified in computers. Family stories and stories of individuals. Stories about me and you. You keep the whole secret and can call it up whenever you want."' (p. 317).

New genetic science allows a revised understanding of the past and the present. Cold cases are solved and new leads developed out of changed understanding. Modern investigation might be able to discover new things about the past using genetics, but similarly new information in the present might bring corruption from the past into the contemporary moment. Indriðason's novel concerns the possible impact of genetic testing on communities, and as such participates in the crime genre's rendering the fears of society concrete. It also demonstrates the new possibilities provided by DNA evidence of investigating crime, and hence of revelation. With the growth of genetic knowledge and the banking of data, the use of DNA evidence is expanding hugely. The ethical decision is about how to maintain this information and to regulate access to it, but also to educate those who would use it to tell 'stories'. Who gets to tell these stories, and their reasons for doing so, is important and reveals an ethics of the revelation of the past. In detective fiction the investigator of the past – normally the detective – takes on the moral role of 'testifying' or witnessing events. In Indriðason's novel that power is taken by the geneticist, who hides rather than reveals. Indeed Erlendur above suggests that the archive itself might be continually telling stories, and these narratives have themselves been created and bought into being by the collection of this information where once they had been buried and unknown. The geneticist, in their collection of data, creates new stories, new histories, and the possibility of new secrets and new interpretations. In *Jar City*, then, enhanced knowledge of the human through genetic investigation leads to a shift in the way that the past is accessed, understood, and known. Indriðason's response to the burgeoning genetic knowledge and the way that it might reconfigure the past articulates a sense of possibility and of complexity. Like many of the moments considered in this book, *Jar City* articulates layers of anxiety about the new genetic information, but also of possibility and even, at times, of reconciliation and redemption.

Critique of Normative Historical Knowledge

As the Dead Sea Scrolls example illustrates, the investigation of genetic material from archaeological sites has begun to generate new data with which to investigate past events. Genetic techniques can enable us to investigate new histories and enrich what is already known. Data derived from material remains and ancient DNA, for instance, has transformed archaeology in the past decade. Increasingly this type of work is no longer confined to prehistory, as increasingly ancient DNA techniques are being used to enable work in more 'modern' historical contexts ranging from medieval pathogens to the gut health of fourteenth-century Jerusalem and Riga to early modern bear-baiting practices.[70] Geneticists are therefore quickly expanding the potential data available for understanding the past, and hence contributing to new ways of understanding and reading that past. In many instances this involves bringing a type of knowledge, particularly about the body, that had previously not been available or considered. Consequently, a different history might be written, one with a new inflection or focus. Rather than a focus on textual sources DNA contributes to a type of history that understands the past through the body.

Public interest and debate about this historical knowledge derived from genetic science is demonstrated by the huge amount of press articles devoted to ancient DNA study over the past five years. In the three months around the writing of this introduction (May-July 2020), *The Guardian* published articles on the Dead Sea Scrolls, Polynesian DNA, Neanderthal DNA, and Viking DNA.[71] In the same period *The New York Times* published articles on the relation between Polynesian and Ancient Native American DNA, the DNA of Neolithic sled dogs, and Stone Age Irish DNA.[72] Each of these articles uses the language of revelation ('sheds light', 'uncovered', 'reveal', 'clues'). Each suggests the importance of DNA work to developing and expanding our understanding of the past, from challenging standard accounts of migration to reconfiguring our ideas about pathogen development. *Homo Sapiens* is no longer the only human of interest, or with a history, as new subspecies are literally unearthed (Denisovans) and more becomes known about those previously thought unimportant (Neanderthals). The intervention of the geneticist into the realm of the historian and the anthropologist is increasingly clear, with, for instance, the *New York Times* using the article on Irish data to suggest that 'Now, genetic data may help delineate social structures of specific communities'. Partly this expansion in coverage is simply due to the explosion in articles about Ancient DNA findings in high-profile scientific journals such as *Nature*. Partly, though, it is part of a wider discourse trying to understand and interpret the human using DNA information. Whether the data in the articles is correctly interpreted or not, the implication of the wealth of press in this area over the past five years is clear: we can begin to tell new stories and uncover important new information through this technology. Indeed, we can recast our historical understanding entirely. Here, genetic interpretation bleeds into historical analysis. On the one hand, new genetic information *reveals* new knowledge; on the other, the techniques of ancient DNA analysis *become* a mode for investigating the past, that is, an historical practice.

This latter issue, the way in which DNA has become linked in the popular historical imagination with revelation and new ways of understanding the past, is key. Genetic information enables a different story to be told. In some instances, such data simply opens up new lines of enquiry. In others it can make a profound shift possible. Work on the DNA of African diaspora, discussed in Chapter 2, demonstrates how genetic data can provide connections long thought lost and buried.[73] In this way genetic information can provide a critique of normative historical knowledge, challenging the centrality of particular types of evidence and ways of narrating the past. The information provided by genetic analysis can become in this way revisionary, insofar as it changes the narrative, but also revolutionary, inasmuch as it significantly challenges and at times overturns standard historical assumptions. The intercession of genetic information has the potential to reorder the way that we think about the past, and the ways in which we understand that past to work. In particular, genetic data can challenge the centrality of the modern concepts of the 'archive' and 'evidence' to historical understanding. As Henry Louis Gates, Jr., suggests about his project to use genetics for genealogical purposes, 'when the paper trail would end, as it inevitably did, in the horrid darkness of slavery, we traced our African roots through our DNA'.[74] Important in Gates, Jr.'s work is the concept of 'reclamation', of grasping something that had been forcibly taken away. In this example, and many others discussed throughout this book, we can see how DNA is seen as a challenge to the accepted 'archive' and 'evidence'. Indeed genetics offer an alternative way of understanding the past, a transgressive and radical way of directly accessing a history that had been intentionally erased.

Understanding of the past through genetics does not necessarily have to be defined as 'postgenomic'; much engagement with genetics has little or nothing to do with contemporary genomic science, or with concepts of periodisation. Throughout, then, I explore the multiple ways that DNA is imagined, how it manifests in the popular historical imagination, and the gap between the 'actuality' and that dreamed metaphor. This consideration allows us to see how governing metaphors affect conception of the past. In Ali Smith's novel *Winter* new genetic knowledge is used to further political arguments against social and cultural division: 'Depends whether you think there's a them and an us, Iris says, or just an us. Given that DNA's let us know we're all pretty much family'.[75] In this instance genetics becomes shorthand for 'species', a way of suggesting a consonance with the rest of humanity and proffering an ideal of solidarity and connection. Iris offers this as a way of short-circuiting debate about migration and suggesting that, in disavowing the historical 'differences' which have kept us apart, that genetics presents a new modernity and the possibility of a better future. This idealistic and future-focused sense of the way that genetics might work socially is often used by genetic data companies (as discussed in Chapter 6). This 'enfranchising' aspect of genetics is deeply problematic, and part of what Jenny Reardon calls 'genomic liberalism'. Indigenous peoples and ethnic minority communities have rightly challenged this sense of the potential of DNA for enfranchisement, pointing out how genetic information has been part of new biocolonial practices.[76] Yet DNA data can produce new readings of the present and the past, expanding our understanding and enabling communities. DNA has the *potential* to enfranchise communities of all kinds in regard to their past, giving them the tools to interpret and understand their own histories. For native and indigenous peoples, DNA might enable a challenge to normative historical assumptions (if data is gathered ethically and in partnership, or, better, by indigenous geneticists). DNA-derived narratives can disavow contemporary biocolonialism and provide a means of articulating a past that has often been denied by the intercession of racist science or historical practice.[77] For other communities DNA presents ways to circumvent standard historical narratives. Iris's comment, then, does suggest a way that the implications of DNA research might be used, actively, to demand a reassessment of social, cultural, and historical assumptions. However this idealistic, utopic version of a past allowing a different future needs to be worked at, and it is also crucial to recognise the sheer power of DNA narratives that tend towards a type of conservative genetic fatalism.[78]

Double Helix History

Throughout this book I consider genetic information from multiple points of view in relation to the understanding, practice, and imagination of history. The book investigates the ways that genetic awareness and practice is seemingly changing historical practice and conceptualisation. For me, 'Double Helix History' is multiple, linking ways of knowing the past with imaginative construction of that past, offering a means for critical interrogation of key historiographical concepts such as evidence, archive, and ethics, whilst at the same time providing a way of reimagining the ways that we conceive of ourselves and aestheticise the past. As with my previous two books on the popular historical imagination, the innovation of the book is to put key historical assumptions under pressure and to attempt to articulate a new way of thinking about how we construct and interact with 'history'. Across my chapters, each investigating a particular aspect of genetic history, I engage with multiple ways of knowing the past. In some parts the geneticist is detective, in others they are part of public memory, in others they offer a means to develop a new narrative. Linking six concepts – 'Public', 'Practice', 'Ethics', 'Politics', 'Self' and 'Imagination – I outline the ways that genetic information, being postgenomic, the public life of DNA, and the genetic historical imaginary work on the body, on collective memory, on the historical imagination, on the ethics of historical investigation, on the articulation of history, and on the collection and interpretation of data regarding the 'past'. In Chapter 1, 'Public', I look at the ways in which DNA as a means for understanding the past is imagined by considering the public articulation of genetics in multiple contexts. The chapter begins to present my argument that data derived from genetic investigation can shift historical understanding and awareness. This discussion leads to Chapter 2, 'Practice', which looks in detail at work on Ancient DNA in order to show how genetic data provides a challenge to historical practice, and, indeed, seems to suggest new ways for knowing and narrating the past. These two chapters dovetail to make an argument about the changes wrought upon historical knowledge and historical investigation by genetic science. Chapter 3, 'Ethics', places such seemingly innovative approaches to the past under scrutiny to understand the ethical implications of considering DNA data as historical 'evidence', and the consequences therefore for history making. This chapter is important in my articulation of a critical historiography of genetic approaches to the past. Chapter 4, 'Politics', continues this critique by investigating contemporary biocolonialism and the way in which genetic approaches to the past have enabled the 'return' of racialised science. Genetic knowledge has the capacity to challenge normative understanding of identity and selfhood in the contemporary and past moments, something discussed in Chapter 5, 'Self' which looks at the phenomenon of genetic genealogy. Through analysing the ways genetic information changes users' understanding of their relation to 'family' past, the chapter contributes to the book's wider purpose in illustrating how DNA influences the ways we think about, engage with, and narrate history. Finally, Chapter 6, 'Imagination' closes the book by looking at the ways in which rappers, artists, poets, novelists, and game creators provide challenges to genetic historical knowledge, seeking ambiguity and uncanny strangeness rather than order and clarity. The book encompasses a wide range of texts, places, contexts, and concepts, and this scope demonstrates the complex ways that genetics pervades our historical imagination and understanding.

Notes

1 Sarit Anava et al, 'Illuminating Genetic Mysteries of the Dead Sea Scrolls', *Cell* 181 (2020), 1-14.
2 Quoted in Josie Glausiusz, 'Ancient DNA Yields New Clues to Dead Sea Scrolls', *Scientific American* 2 June 2020, https://www.scientificamerican.com/article/ancient-dna-yields-new-clues-to-dead-sea-scrolls/ [accessed 19 June 2020].
3 Donna Harraway, *Modest Witness* (London and New York: 1997), Paul Rabinow, *French DNA* (Chicago, IL: University of Chicago Press, 2002), Jenny Reardon, *Race to*

the Finish: Identity and Governance in an age of genomics (Oxford: Oxford University Press, 2005), Kim TallBear, *Native American DNA* (Minneapolis, MN: University of Minnesota Press, 2013), Josie Gill, *Biofictions* (London: Bloomsbury, 2020), Catherine Nash, *Genetic Geographies* (Minneapolis, MN: University of Minnesota Press, 2015), Judith Roof, *The Poetics of DNA* (Minneapolis, MN: University of Minesota Press, 2007), Clare Hanson, *Genetics and the Literary Imagination* (Oxford: Oxford University Press, 2020), Barry Barnes and John Dupré, *Genomes and what to make of them* (Chicago, IL: University of Chicago Press, 2008).

4 Jackie Stacey, *The Cinematic Life of the Gene* (Durham, NC: Duke University Press, 2010) p. xii; Alondra Nelson, *The Social Life of DNA* (Boston, MA: Beacon Press, 2017).

5 Nadia Abu El-Haj, *The Genealogical Science* (Chicago UP, 2012), p. 22.

6 The historical movement across the twentieth century to this point is traced by Sommer in *History Within*.

7 Two books that do undertake this although in different ways to the present book are Christine Kenneally, *The Invisible History of the Human Race* (New York: Viking Penguin, 2014) and Marianne Sommer, *History Within* (Chicago, IL: University of Chicago Press, 2016).

8 See 'Introduction: Genetic Claims and the Unsettled Past' in *Genetics and the Unsettled Past* ed. Keith Wailoo, Alondra Nelson, and Catherine Lee (New Brunswick, NJ: Rutgers University Press, 2012), pp. 1-13 and Pramod K. Nayar 'Autobiogenography: Genomes and Life Writing', *a/b: Auto/Biography Studies* 31:3 (2016), 509-25.

9 *A Brief History of Everyone Who Ever Lived* (London: 2017), p. 4.

10 'Preface' in Chris Pomery, *DNA and Family History* (PRO Publications: 2004), p. iv.

11 Wolfgang Haak et al, 'Massive migration from the steppe was a source for Indo-European languages in Europe', *Nature* 522 (2015), 207-211 (207).

12 This is clearly not the only type of new data that is changing the way that 'history' works. Other new information that challenges historical normativity might relate to climate crisis, 'big' history, synthetic information, Indigenous historiography, biohistory, the Anthropocene, see for instance Marnie Hughes-Warrington and Anna Martin, *Big and little histories* (London and New York: Routledge, 2021).

13 See Daniel Woolf, 'Getting back to normal: On normativity in history and historiography', *History and Theory* 60:3 (2021), 469-512.

14 See Siddharta Mukherjee, *The Gene: An Intimate History* (London: Vintage, 2017), Sommers, *passim*, and Nikolai Krementsov, *International Science between the World Wars: The Case of Genetics* (London and New York: Routledge, 2004).

15 'Molecular Structure of Nucleic Acids: A Structure for Deoxyribose Nucleic Acid', *Nature* 171 (1953), 737-8.

16 Outlined in Georgina Ferry, 'The Structure of DNA', *Nature* 575 (2019), 35-6.

17 Mukherjee, *The Gene*, p. 13.

18 'DNA Sequencing with chain-terminating inhibitors', *PNAS* 74:12 (1977), 5463-7.

19 Frederick Sanger and A.R. Coulson, 'A rapid method for determining sequences in DNA by primed synthesis with DNA polymerase', *Journal of Molecular Biology* 94:3 (1975), 441-8; David A. Jackson, Robert H. Symons, and Paul Berg, 'Biochemical Method for Inserting New Genetic information into DNA of Simian Virus 40', *PNAS* 69:10 (1972), 2904-09; 'The Nobel Prize in Chemistry 1980', https://www. nobelprize.org/prizes/chemistry/1980/summary/ [accessed 9 July 2021].

20 Sahotra Sarkar, *Genetics and Reductionism* (Cambridge: Cambridge University Press, 1998) and Stacey, *The Cinematic Life of the Gene*.

21 *Narrative in the Age of the Genome*, p. 21.

22 Robert L. Sinsheimer, 'The Santa Cruz Workshop – May 1985', *Genomics* 5:4 (1989), 954-6.

23 Robert Cooke-Deegan, *The Gene Wars: Sciences, Politics, and the Human Genome Project* (London and New York: Norton, 1995).

24 Francis S. Collins et al, 'New Goals for the U.S. Human Genome Project, 1998-2003', *Science* 282: 5389 (1998), 682-9.

25 'All Goals Achieved', NHGRI press release, 14 April 2003, archived at https://www.genome.gov/11006929/2003-release-international-consortium-completes-hgp [accessed 15 June 2021].

26 NHGRI Press Release, 14 April 2003, held at https://www.genome.gov/11006929/2003-release-international-consortium-completes-hgp [accessed 19 August 2020].

27 'Joint Proclamation', April 2003, archived at https://www.genome.gov/sites/default/files/media/files/2021-02/2003_Joint_Proclamation.pdf [accessed 15 June 2021].

28 Sara Reardon, 'A complete human genome sequence is close', *Nature* 594 (2021), 158-9.

29 Aída Falcón de Vargas, 'The Human Genome Project and its importance in clinical medicine', *International Congress Series: Current Trends in Clinical Medicine* 1237 (2002), 3-13.

30 Richard A. Gibbs, 'The Human Genome Project changed everything', *Nature Reviews Genetics* 2020, https://doi.org/10.1038/s41576-020-0275-3.

31 Wilhelm J. Ansorge, 'Next-generation DNA sequencing techniques', *New Biotechnology* 25:4 (2009), 195-203 and Elaine R. Mardis, 'The impact of next-generation sequencing technology on genetics', *Trends in Genetics* 24:3 (2008), 142-9.

32 'The Human Genome Project: Past, Present, and Future', *Science* 248: 4951 (1990), 44-9 (p. 49).

33 Jenny Andersson, *The Future of the World: Futurology, Futurists, and the Struggle for the Post-Cold War Imagination* (Oxford: Oxford University Press, 2018).

34 Human Genome Program, US Department of Energy, *Human Genome News* 7:3 (1995), p. 1.

35 Francis S. Collins, Michael Morgan, Aristides Patrinos, 'The Human Genome Project: Lessons from Large-Scale Biology', *Science* 300: 5617 (2003), 286-90.

36 *Genome* (London: Harper Perennial, 1999), p. 5.

37 See Ludmilla Jordanova, *Nature Displayed: Gender, Science and Medicine 1760– 1820* (London: Longman, 1999) and Londa Schiebinger, *Nature's Body: Gender and the making of modern science* (Boston: Beacon Press, 1993).

38 This locational desire has long been associated with genetic science, see the essays collected in Jean-Paul Gaudilière and Hans-Jörg Rheinberger, *From Molecular Genetics to Genomics: The mapping cultures of 20th-century genetics* (London and New York: Routledge, 2004).

39 Victor K. McElheny, *Drawing the Map of Life* (New York, N.Y.: Basic Books, 2010).

40 Donna Haraway, *Modest_Witness@Second_Millennium.FemaleMan©_Meets_OncoMouse™: Feminism and Technoscience* (London and New York: Routledge, 1997).

41 See the discussion of space and genetics in Gísli Pálsson, *Anthropology and the New Genetics* (Cambridge: Cambridge University Press, 2007) and Catherine Nash, *Genetic Geographies*.

42 'President Clinton announces the completion of the first survey of the entire human genome', White House Press Release, 25 June 2000, in the Human Genome Project Information Archive 1990-2003, https://web.ornl.gov/sci/techresources/Human_Genome/project/clinton1.shtml [accessed 6 August 2020].

43 https://www.genome.gov/human-genome-project [accessed 19 August 2020].

44 See El-Haj, *Genealogical Science*, p. 13, and Reardon, *passim*.

45 Choksey, p. 13.

46 This is, of course, hardly new in rhetoric around scientific innovation, see for instance Steven Shapin, *The Scientific Revolution* (Chicago, IL: University of Chicago Press, 1994) and Charlotte Sleigh, *Literature and Science* (Basingstoke: Palgrave Macmillan, 2011).

47 Hallam Stevens and Sarah S. Richardson, 'Beyond the Genome' in *Postgenomics: Perspectives on Biology After the Genome*, eds. Sarah S. Richardson and Hallam Stevens (Durham and London: Duke University Press, 2015), pp. 1-8.

48 See Hans-Jörg Rheinberger and Staffan Müller-Wille, *The Gene: from Genetics to Postgenomics*, trans Adam Bostanci (Chicago, IL: University of Chicago Press, 2018) and the more popular book by Joshua Z. Rappoport, *Mapping Humanity: how modern genetics is changing criminal justice, personalized medicine, and our identities* (Dallas, TX: BenBella, 2020).

49 Benjamin R. Bates, 'Public culture and public understanding of genetics: a focus group study', *Public Understanding of Science* 14:1 (2005), 47-65.

50 Jay Clayton, 'Genome Time', in *Time and the Literary* ed. Karen Newman, Jay Clayton, and Marianne Hirsch (New York: Routledge, 2002), pp. 31-59 (p. 31).

51 Nikolas Rose, *The Politics of Life Itself* (Princeton, N.J.: Princeton University Press, 2007), p. 13.

52 Paul Rabinow, *French DNA*, p. 9.

53 Paul Gilroy, *Against Race*, p. 37.

54 Kaushik Sunder Rajan, *Biocapital: The Constitution of Postgenomic Life* (Durham, N.C.: Duke University Press, 2000).

55 Jenny Reardon, *The Postgenomic Condition: Ethics, Justice and Knowledge After the Genome* (Chicago and London: University of Chicago Press, 2017), p. 17.

56 Jackie Stacey, *The Cinematic Life of the Gene* (Durham, N.C.: Duke University Press, 2010).

57 Lara Choksey, *Narrative in the Age of the Genome* (London: Bloomsbury, 2021).

58 Martyn Pickersgill, Jörg Niewöhner, Ruth Müller, Paul Martin, and Sarah Cunningham-Burley, 'Mapping the new molecular landscape: social dimensions of epigenetics', *New Genetics and Society* 32:4 (2013), 429-47 (429). On epigenetics see also Miranda R. Waggoner and Tobias Uller, 'Epigenetic determinism in science and society', *New Genetics and Society* 34:2 (2015), 177-95. The continuing discussion of genetics and society is shown by the number of new publications in this area, including Gill, *Biofictions*, Choksey, *Narrative in the Age of the Genome*, the *Medical Humanities* Special Edition 'Global Genetic Fictions' ed. Clare Barker, 47:2 (2021), and *Game of Bones: Critical Perspectives on Ancient DNA* ed. Anna Källén, Charlotte Mulcare, and Daniel Strand (Cambridge: Cambridge University Press, forthcoming 2022).

59 *Poetry* June 2003, available at https://www.poetryfoundation.org/poetrymagazine/poems/41869/mapping-the-genome, ll. 1-3.

60 Samuel Y. Edgerton, *The Mirror, the Window, and the Telescope: How Renaissance Linear Perspective Changed our Vision of the Universe* (Ithaca, NY: Cornell University Press, 2009).

61 Michael L. Blakey, 'Bioarcheology of the African Diaspora in the Americas: Its Origins and Scope', *Annual Review of Anthropology* 30 (2001), 387-422 and Margrit Shildrick, 'Genetics, Normativity, and Ethics: Some Bioethical Concerns', *Feminist Theory* 5:2 (2004), 149-65.

62 Sylvia Wynter and Katherine McKittrick, "Unparalleled Catastrophe for Our Species? Or, to Give Humanness a Different Future: Conversations" in *Sylvia Wynter: On Being Human As Praxis*, ed. Katherine McKittrick (Durham, NC: Duke University Press, 2014), 9-90 (p. 20).

63 Pramod K. Nayar 'Autobiogenography: Genomes and Life Writing', *a/b: Auto/Biography Studies* 31:3 (2016), 509-25.

64 Niklas Rose, 'The Human Sciences in a Biological Age', *Theory, Culture & Society* 30: 1 (2012), 3-34 (p. 25).

65 David Reich in Sarah Zhang, 'Ancient DNA is Rewriting Human (and Neanderthal) History', *The Atlantic*, 14 March 2018, https://goo.gl/LEVsEv [accessed 27 March 2018].

66 *Spectres of Marx*, p. 6.

67 Andrew Hui, 'Horatio's Philosophy in *Hamlet*', *Renaissance Drama* 41: 1/2 (2013), 151-71.

68 The Icelandic population is small, homogenous, and relatively immobile, so it has been subject to a great deal of state and private genetic testing, see Gísli Pálsson and Paul

Rabinow, 'Iceland: the case of a national human genome project', *Anthropology Today* 15:5 (1999), 14-18.

69 Arnaldur Indriðason, *Jar City*, trans. Bernard Scudder (London: Vintage, 2010; first published in Iceland in 2000).

70 Verena J. Schuenemann et al, 'Targeted enrichment of ancient pathogens yielding the pPCP1 plasmid of *Yersinia pestis* from victims of the Black Death', *PNAS* 108: 38 (2011), E746-52; Susanna Sabin et al, 'Estimating molecular preservation of the intestinal microbiome via metagenomic analyses of latrine sediments from two medieval cities', *Philosophical Transactions of the Royal Society of London, Series B: Biological Sciences* 375: 1812 (2020), 1-15; 'Box Office Bears' AHRC-funded project described at https://beforeshakespeare.com/2020/08/03/box-office-bears-a-new-research-project-on-animal-baiting/ [accessed 2 July 2021];

71 Nicola Davis, 'Ancient DNA is offering clues to puzzle of Dead Sea Scrolls, say experts', *The Guardian* 2 June 2020, https://www.theguardian.com/science/2020/jun/02/ancient-dna-helps-experts-tackle-puzzle-of-dead-sea-scrolls; Agencies, 'Indigenous Americans had contact with Polynesians 800 years ago', *The Guardian* 8 July 2020, https://www.theguardian.com/world/2020/jul/08/indigenous-americans-polynesians-dna-800-years-ago; Nicola Davis, 'Humans and Neanderthals 'co-existed in Europe for longer than is thought'', *The Guardian* 11 May 2020, https://www.theguardian.com/science/2020/may/11/humans-and-neanderthals-co-existed-in-europe-far-longer-than-thought; Nicola Davis, 'Researchers find earliest confirmed case of smallpox', *The Guardian* 23 July 2020, https://www.theguardian.com/science/2020/jul/23/researchers-find-earliest-confirmed-case-smallpox-viking-era [all accessed 6 August 2020].

72 Carl Zimmer, 'Some Polynesians carry DNA of Ancient Native American', *New York Times* 8 July 2020, https://www.nytimes.com/2020/07/08/science/polynesian-ancestry.html?searchResultPosition=1; James Gorman, 'Dog Breeding in the Neolithic Age', *New York Times* 25 June 2020, https://www.nytimes.com/2020/06/25/science/arctic-sled-dogs-genetics.html?searchResultPosition=3; James Gorman, 'DNA of 'Irish Pharaohs' Sheds Light on Ancient Tomb Builders', *New York Times* 17 June 2020, https://www.nytimes.com/2020/06/17/science/irish-archaeology-incest-tomb.html?searchResultPosition=4 [all accessed 6 August 2020].

73 See Alondra Nelson, *The Social Life of DNA*.

74 *In Search of Our Roots: How 19 Extraordinary African Americans Reclaimed their Past* (New York: Random House, 2009), p. 11.

75 Ali Smith, *Winter* (London: Hamish Hamilton, 2017), p. 206.

76 Laurie Anne Whitt, 'Biocolonialism and the commodification of knowledge', *Science as Culture* 7 (1998), 33-67.

77 Krystal Tsosie, Rene L. Begay, Keolu Fox, Nanibaa' A. Garrison, 'Generations of genomes: advances in paleogenomics technology and engagement for Indigenous people of the Americas', *Current Opinion in Genetics and Development* 62 (2020), 91-6.

78 I owe this phrase to Gerben Bakker.

1

PUBLIC

This chapter explores how genetics and concepts of the past are combined in the wider historical imagination by focusing on the public iteration of DNA. The chapter considers the 'public history' of DNA, that is, the ways in which the molecule manifests in the historical imaginary and contributes to wider understanding of the past. Some of the most high profile public events of the past two decades relating to DNA and genetics had an historical aspect: the discovery and reinternment of Richard III; the revelation of Thomas Jefferson's relationship with his slave, Sally Hemings; the genetic database of the African Burial Ground project; Henry Louis Gates, Jr's use of DNA in his genealogical television programmes; the use of genetic technologies to investigation historical war crimes and genocide in Argentina, Mexico, Rwanda, and the former Yugoslavia. These cases, and others, have served to intertwine genetics and historical investigation in the public imagination, and DNA has become associated with understanding the past, to the extent that an entire fictional genre (cold cases) has evolved to reflect this. DNA has changed the way the historical imagination works, contributing a new, material aspect to investigation of the past. This chapter, then, explores several of these examples of the 'public life' of DNA in order to illustrate the contexts for discussion of genetic knowledge, as well as to conceptualise the complex interrelationship between genetics and access to the past. Each section considers in detail the ways in which public memory intertwines with the genetic imagination to create communal practices of recollection, imagination, and interrogation.

DNA and the double helix present a case-study for thinking about public historical issues post-2000, particularly with relation to commemorative practice, heritage, display, and the body. In this sense, the public historical function of DNA can be seen to be a fundamental and foundational slippage between then and now, real and imagined. The examples discussed in this chapter illustrate the tension in the public imagination between DNA as something material and 'real',

DOI: 10.4324/9781003052975-2

and DNA as an imagined concept. They allow us to consider the contexts in which ideas relating genetics will be received and understood. Telling the story of (and with) DNA involves an intervention into the spaces of public history. The way that we imagine DNA and the way that its history is told demonstrates something about the way that we choose to remember as a society.

Additionally, as is made evident throughout this chapter, DNA and the double helix also stand as a motif for accessing the past, a particular way of engaging with historical material and telling historical narrative. In particular, this chapter shows how DNA can participate in discourses of revisionism and rethinking, introducing as a motif an evidentiary shift and novelty of historical understanding. This chapter discusses genetics *in* public history – that is, the ways in which DNA has been commemorated and what this suggests for how and why we remember – and also genetics *as* public history – that is, as a motif for thinking about the way that memory is formed and how it might develop and change. Genetic information can challenge received historical discourse, as in the case of Sally Hemings, but it can also transform the way we might think about our relationship to the past entirely. Genetic information can create its own archive, as in the case of James Watson's genome sequence, but it can also suggest a way of seeing the past, as is the case in the public memory of Rosalind Franklin. DNA is a revisionist tool, challenging normative assumptions and allowing a new light to be shone. As can be seen in the discussion of the chapter about the work of novelist Ali Smith, DNA can offer an alternative to a particular type of history, provide a new means for approaching the past and conceptualising history. In its public iterations, then, DNA flickers as real and not-real, solid and imagined, small and large; it can challenge our sense of scale, of time, and of self. In the examples that are discussed through this chapter, we see the public life of DNA, something unknown and banal, highly recognisable and barely seen, uncanny, and familiar. Considering the public life of DNA, then, allows an understanding of the interrelations between DNA and 'history', this latter as something practiced, imagined, and constructed.

Public Knowledge of Genetics and COVID-19 Vaccines

Whilst genetic understanding was relatively wide before the 1990s, there has been an acceleration in usage.[1] Lay genetic knowledge – the 'double helix', the 'building blocks of life', genetic inheritance, and skewed ideas of race – is widespread, demonstrated by the proliferation of popular science books on the subject.[2] The surge in interest and understanding of DNA post-2000, then, provides a context for investigating the ways in which genetics has complicated our understanding of the past. Wider awareness of genomics and DNA has to a certain extent merely cemented in the public imagination particular modes of usage and governing metaphors. A few recent examples can show the popular prevalence of genetic understanding, and the diverse ways that genetic thinking is embedded into cultural consciousness so much it is barely noticeable. Firstly, 'England DNA' is a project undertaken by the English Football Association to promote the 'coaching and playing philosophy of the England Teams'.[3] This is an attempt to impose a template on sport through practice and ethos, using 'DNA' as a shorthand for inherited characteristics and the building blocks of an identity. This usage isn't unusual in discussing sport, as examples such as 'the Barça DNA', 'Formula One's DNA', and 'The community game is the DNA of the sport [Rugby]' demonstrate.[4] The sense of genetics as something foundational is widespread in the usage of the metaphor and has hardened in the past two decades. Writing in summer 2020 the *New York Times* on conservatism and Donald Trump, Peter Wehner argued that 'because of its reverence for our Madisonian system of government – checks and balances, separation of powers – conservatism considers compromise part of our constitutional DNA'.[5] Similarly the British MP Nadhim Zahawi claimed in September 2020 that 'It goes against the grain, the DNA of a Conservative government, to curtail people's liberties'.[6] Critic Eric Torres, re-calling the musician Sylvester, argued '"You Make Me Feel" pumps with the same space-age DNA as Donna Summer's "I Feel Love," released just the year before'.[7] This use of 'DNA' as referent for something fundamental is commonplace in public discourse. A further example can be seen in the online phenomena linking the 'building blocks of life' concept to literal building blocks. There exist online images of hundreds of LEGO models of DNA Including 'Mr DNA', the explanatory cartoon character from *Jurassic Park*, rendered in LEGO.[8] Materials scientists have reversed this by using LEGO as a means to model DNA structures.[9] Here the association of DNA with foundation is key, as LEGO enables an understanding of genetic structure as architectural and three-dimensional]. A final example is the huge success of the *Jurassic Park* films (six films to date, 1993-, discussed in Chapter 3), exploring a fear of genetic engineering and the commodification of genes. These films communicate the sense of DNA as foundational and manipulable. There are countless other moments which illustrate the widespread popular understanding of DNA, regularly reduced to certain tropes and metaphors (see further discussion in Chapter 6). Public understanding of genetics is widespread if imperfect.

We can understand the complexity of public genetic knowledge by looking at the development of vaccines to combat the global COVID-19 pandemic of 2020-22. The emergence of COVID-19 in late 2019 led to millions of deaths, a worldwide shutdown, and a 'race' for a vaccine as an escape from the threat of the virus. A number of vaccines were developed, some using innovative gene-based techniques.[10] Gene-based vaccine candidates were different to standard vaccination techniques insofar as rather than traditional methods they used new genetic approaches using messenger RNA (mRNA) to 'teach our cells how to make a protein' that would force an immune response.[11] This type of vaccine was considered preferable because of its 'rapid and low-cost manufacturing process'.[12] A huge effort was mobilised to sequence the genome of the virus and its mutations.[13] The mRNA vaccine candidates worked well and were quickly developed for usage worldwide.[14] The company Pfizer/ BioNTech, whose mRNA vaccine was developed at great speed, were at the forefront of efforts and their product was the first in the world to be used, licenced in the United Kingdom in December 2020 (most vaccines take 10 years to licence).[15] Another company, Moderna, was licenced in January 2021 with a vaccine that the UK government explained 'works by injecting a small part of the COVID-19 virus' genetic code, which triggers an immune response and creates antibodies in the human body able to fight the virus'.[16]

The shift to mRNA was heralded as a 'a huge leap forward for the science of vaccine making' and seemed to suggest new ways that genetic-led medicine might work in the future.[17] The widespread reporting on the genetic work that the Pfizer/BioNTech and Moderna vaccines undertake demonstrated the turbocharging of genetic medicine. Genetic knowledge and practice was harnessed to meet the greatest public health challenge of the past 50 years. However, the development and deployment of genetically-based vaccines led also to online panic and conspiracy about how they would work and what they would do to the individual taking them. In particular stories began to circulate falsely suggesting that mRNA vaccines alter the DNA of the recipient.[18] Articles and video clips shared on social media also wrongly claimed Bill Gates had said that new vaccines were designed to 'alter' and 'manipulate DNA'.[19] Anti-vaccination groups and online conspiracy theorists shared multiple false reports about Gates and the mRNA vaccine around the world.[20] Additionally, particular concerns were expressed by the Catholic and the Jewish communities and doctors recognised that they would need to respond to various rumours about the effects of the vaccine.[21] These anxieties about DNA and the COVID-19 vaccine demonstrate how important and foundational the concepts of genetic identity have become. They show how medical conspiracy theory can quickly warp public debate. They also illustrate that wider public understanding of DNA is at once fragile and robust. Fragile, that is, because public genetic knowledge is predicated upon incomplete understanding of highly complex issues, and easily manipulated. Robust, because the link in the public imagination between DNA and 'self' is incredibly strong. The concerns that mRNA vaccines might change a recipient's DNA are to a certain extent of a piece with wider public debates about the precision of genetics in determining identity and behaviour.

Public Commemoration and Revision

Bruno J Strasser suggests that the use of the double helix model can be seen to 'escalate' in the 1990s with the move towards mapping the genome.[22] Asking how the model became so important and iconic in our collective memory, Strasser shows how citation of Crick and Watson's 1953 paper creates a 'genealogy' of knowledge. This understanding of the development of knowledge as something with hierarchical and familial structure conceives of science as something developmental and, to a certain extent, linear. In this section, I show how the association of DNA with Crick and Watson is part of a wider cultural memory of scientific achievement that is being slowly challenged. In this sense the creation of a 'genealogy' within the historical imagination is important as it shows how society has come to remember and the things that it has chosen to highlight. 'Double Helix History' here is the history of the memory of the double helix, and what this implies for national memory structures and the way that science figures in the contemporary historical imagination.

Commemorating Genetics

On 27 February 2003, the 108th United States Congress passed a concurrent resolution sponsored by members of both the Senate and Congress:

> Designates: (1) April 25, 2003, as DNA Day in celebration of the 50th anniversary of the publication of the description of the double-helix structure of DNA (deoxyribonucleic acid) by James D. Watson and Francis H.C. Crick; and (2) April 2003 as Human Genome Month to celebrate that anniversary, the essential completion of the sequence of the human genome, and the development of a plan for the future of genomics.[23]

This resolution recommended that the day be recognised and celebrated by 'schools, museums, cultural organisations and other educational institutions across the nation' to recognise 'one of the most significant scientific discoveries of the 20th century'. The date of April 2003 was also claimed as the formal completion of the Human Genome Project, and the resolution therefore commemorated past achievement as well as contemporary. Subsequently DNA Day has been celebrated by the National Human Genome Research Institute in the USA and to a lesser extent around the world on 25 April.[24]

The announcement of DNA Day signified the accession of genetics into a commemorative calendar, the work becoming part of American public heritage culture. 'DNA Day' establishes genetics as something worthy of remembering and commemorating, a national celebration of human achievement and scientific development. The commemorating of the double helix description date seeks to anoint an originary moment for the understanding of DNA. The anniversary is celebrated by looking to work in the future, and the description of the double helix is seen as the foundation for current research. Significant here is the date of the description of the structure, rather than the subsequent work to establish this as correct (although the resolution acknowledges the amount of research that has been undertaken in this area). It suggests a heroic, originary model of achievement and focuses only on Crick and Watson (neglecting Maurice Wilkins, who shared the Nobel Prize with them in 1962, and Rosalind Franklin, whose work was foundational to the work). Hence, the resolution participates in public culture of focusing on the figurehead participants rather than seeing scientific work as collaborative. This particular naming and memory practice is an historical intervention, inviting discussion of what and who is commemorated, why and how. DNA Day therefore figures the history of the description of the double helix as something living and part of the wider cultural memory. Moreover, the Congress resolution gives some instruction as to how this should be done, and *why*. We we can see the structure of commemorative practice expressed.

The double helix, and DNA research more generally, becomes therefore a means of accessing the past and a shorthand for certain achievement. The anniversary will be commemorated and interpreted in public by educational institutions. This signals one way that genetics enters into historical discourse, that is, as subject of and participant in public recollection through commemoration and anniversary activities. Similar 'public' moments are the debate over whether Rosalind Franklin should be on the new £50 note in the United Kingdom during 2018, and the naming of the Francis Crick Institute for Biomedical Research in London in 2011. Both these latter moments see the scientific figures becoming subsumed into commemorative practice. Both were not 'neutral' moments. The discussion about Franklin regularly now paints her as the forgotten and neglected woman of genetic science, sometimes obscuring her actual achievements (discussed below). Putting a woman on bank notes in the United Kingdom has not been an easy thing to do, with MPs receiving death threats after the campaign to put Jane Austen on the £10.[25] In the case of Franklin in 2018 the Bank of England asked the public for suggestions for scientists to appear on the £50, seeking a collaborative process; despite many women appearing in the discussion the BoE eventually chose Alan Turing. In the case of the Crick, it is notable that the Institute does not commemorate in its name the contribution of Watson (there is also a Rosalind Franklin Institute), instead focusing on the one of the pair despite their being linked in the public consciousness. It is the case that Watson is still alive and is now a figure of some controversy due to his political views of science.

Considering the ways in which genetic science has been publicly commemorated entails rethinking the role of history in public. It is important to think about types of memorial practice that have not been regularly considered. For instance, we need to recognise the naming of institutes and laboratories as public historical practice.[26] Often scientific institutions are not considered to be historical organisations. However, as mentioned, there is a Francis Crick Institute in London, a Rosalind Franklin Institute in Oxfordshire, a Rosalind Franklin University in Chicago, and the Wellcome Sanger Institute complex includes the Sulston laboratories. Additionally, there are many smaller buildings named after various key figures in the development of genetic science, as well as numerous prizes and lectureships. Such naming matters, as the controversy around the Francis Galton Lecture theatre at University College London demonstrates. Galton's work on eugenics was destructive, and many called for the theatre to be renamed, which it was in summer 2020.[27] This issue is increasingly high profile and the discussion, given later, of the public memory of James Watson shows that genetic science is not immune from being part of public discussions of commemorative practice and ethics. Naming supposes commemoration and does not allow for a particularly subtle version of the past to be communicated. Science institutions are rarely considered in this way in discussions of public history, and their sites of memory remain relatively uninterrogated. Naming of labs, buildings, prizes, lectureships, and institutes suggests a model of linearity (standing on the shoulders of giants; building on the foundations laid by heroic individuals) and a particular way of remembering within the scientific community and more widely. It also clearly contributes to the historical imaginary, suggesting a mode of re-collection which generally celebrates male, white achievement.

The ways that the 'achievements' of DNA are commemorated participate in a wider discourse of heroic celebration. In this vein of public commemoration then there is a plaque to Crick in Westminster (St. George's Square), one for Watson also in London (Vincent Square), and two at the Eagle pub in Cambridge where they first announced their work (inside and outside). This type of public memorial is important in establishing the heritage significance of particular figures. Crick is described in some detail including 'Discoverer of DNA & the Genetic Code' whereas Watson is simply 'DNA Scientist'. The plaque at the Eagle has been changed with graffiti at certain times to reflect the contribution of Rosalind Franklin. The Eagle plaque tells a reader that 'On this spot' Crick and Watson made their announcement, underlining the key aspect of heritage commemoration of this type, that is, the importance of particular location to memorial practice. In this instance the revelation of particularity and tourist re-enactment through at-tendance at the space is key. This is part of the wider locational memorial practice of the plaques, and the importance of the physical is also see in town planning: there are streets named after Crick in Cambridge, Guildford and Northampton; and Watson formerly in Norwich (see also the Cambridge cycle route, below). This continued public focus on Crick and Watson communicates a particular version of discovery and progress.

As a prominent figure in this memorialisation James Watson can be considered a case-study for how public memory might shift and change. As a 'celebrity' as-sociated with DNA he has been celebrated in various ways although he is also a controversial figure and has been shunned by the science establishment for several decades. After numerous comments involving anti-Semitism, race, intelligence, sexuality, and gender in the early 2000s he became increasingly isolated.[28] A PBS documentary about his work led to Watson having some of his honours stripped from him by Cold Spring Harbor Laboratory in 2019 due to his comments about genetics and race.[29] In July 2020, a road named after Watson in Norwich was changed to be named after Rosalind Franklin.[30] Similarly, in 2019, Halls of Residence at the University of Portsmouth formerly named after Watson were renamed in memory of Franklin. These changes demonstrate the continuing public debate over public memorials and the naming of places, and the ways in which commemorative practice changes according to contemporary under-standing. Watson's 'achievements' remain, what has changed is the public memory of his work and the ways in which society wishes to remember. In this, then, we see that scientists are similar to other political figures about whose public com-memoration there is significant debate.

A high proportion of the articles about Watson's controversial statements refer to him as the 'Father of DNA', conferring him progenitor status and emphasising again the sense of creation and linear hierarchy associated with scientific development and progress.[31] Watson's whole genome was sequenced in 2007, seemingly only the second complete genome to be described in this way.[32] After he was presented with the genome data, Watson made it available for medical research.[33] This was very early in the process of sequencing and the work was undertaken as part of a drive to make sequencing processes quicker and cheaper: 'The formidable size of the diploid human genome, approximately 6 gigabases, has prevented the routine application of sequencing methods to deciphering complete individual human genomes'.[34] The work on Watson's genome therefore was important in directing efforts this way. It was also hugely symbolically significant, and again positions Watson – and in this case, his genetic data – as originator and pioneer. This was a keynote moment after the HGP, and Watson's actions show that the 'commemorative' act of presenting him with his genome might also be a public moment. His data became a reference point, an archive of information. Sequencing of genomes is now extremely common, and hundreds are now online and accessible for researchers to work on. Hence, the body becomes an archive, to be investigated and analysed. This can be extremely problematic, as the discussion of the HGDP in Chapter 3 discusses. The sequencing of Watson's genome was an intervention into historical practice. His genetic data became a statement of his importance and an expression of commemoration. The work was a tribute to what he had done and an expression of the significance of his research; it also underlined the significance of *him* as a body, as a collection of data, and as a scientific progenitor. His genetic archive became a public entity, and 'James Watson' proliferated, being both a person and a set of data.

Watson sold his Nobel Prize medal in 2014.[35] He claimed it was to raise money for himself (due to being *persona non grata*) and for medical research. The medal was bought by the Russian oligarch and owner of Arsenal F.C. Alisher Usmanov for $4.1m; Usmanov immediately returned it to Watson.[36] Watson therefore becomes intertwined with his public profile and his own DNA. He is imaginatively associated with the 'discovery' of DNA and his own genetic code becomes part of his public profile. He is considered 'significant' and public, important, and worthy of remembrance. On the one hand, Watson manifests a public 'Double Helix History' insofar as he is symbolic of the way that DNA is remembered. He is cited as the father, the originator, and the creator of the DNA myth. Watson's public profile also demonstrates how political genetic science is; even his marginalisation has forced discussion of certain attitudes towards race and sexuality. Furthermore, Watson's actions in the past decades have created a bizarre public historical profile, contributing to commemoration culture and archive practice as well as reminding us that artefacts such as prizes are still resonant material things. Finally, the dearth of buildings named after Watson testify to a residual recognition of the powerful nature of naming in the public historical imagination.

Imagining DNA in Public

Whilst the focus on the particular figures communicates a particular aspect of the public memory of DNA, within public commemorative practice the double helix itself has become a widely recognisable icon of commemoration. There are a variety of public art works representing the structure around the world, including sites in Frankfurt, Dayton (Ohio), Burlington (North Carolina), Birmingham, Beijing, Dublin, Moscow, and London. Other public structures include a bridge (Singapore) and a children's play area (Berkeley, California). Charles Jencks created a version of the double helix in for Clare College, Cambridge in 2005; Jencks then developed the visual language for work found in Newcastle, Glasgow, London, and Dublin (Figure 1.1).[37]

Jencks was a key theorist of postmodern architecture and articulator of theories of pluralism and The involvement of Jencks in DNA memorials and artworks suggests that the development of a kind of iconography is not simply coincidence. Key to most of the sculptures that have been mentioned are the fact that they are built in reflective metal, are three-dimensional, and relatively large in size. Jencks's sculpture for Clare College reproduces the hour-glass shape of the double helix but roots this to the ground with layered landscaping at the base even as the helix unravels at the top to create two arms which allow the piece to stand. The material is galvanised steel, both impressively shiny but also evidently non-realistic and

FIGURE 1.1 Clare College Memorial Court

seemingly non-organic. Jencks's work renders DNA as something unworldly, impressive, elegant, and unreal. The combination of materials – man-made and natural – is a postmodern trope referring to eclecticism. Galvanised steel is mainly used in construction, providing the frameworks for buildings. Jencks brings the inner frame outside, focusing on the thing that structures but is usually unseen. He focuses on the paradox of DNA, that is, its ineffable materiality. The sculptures are not realistic but reflect upon their own wroughtness, commenting on the way that genetic material 'lives' in the imagination, both real and unreal. The work is connected to the world, suggesting a way of conceiving genetic material that is both imagined and solid simultaneously. This inbetweenness and simultaneity are again a keynote of the postmodern aesthetic, a conscious echoing of now and then and a refusal to be still. In this way, the modernity of DNA – now, us, cutting-edge, shiny, reflective – is contrasted with the commemorative function of the sculpture. DNA is then and now, pulling back in time but simultaneously in the moment. Postmodern sculpture challenges temporality and in this the re-presentation of DNA becomes something that is atemporal, both futuristic and situated. The meta-commentary of postmodern aesthetics lies in its duality, its ability to be 'both', an irony that can be found in buildings revealing their foundations or sculptures playfully reflecting the image of the viewer. In the case of Jencks's work, the DNA sculptures invert reality, presenting something small a huge scale, and bringing the foundation into the light. This aesthetic of the double helix memorial as something that is simultaneously now/modern and then/timeless contributes to a way of thinking and conceiving of genetic information.

The set of tropes associated with the representation of DNA that Jencks developed is now relatively standard. Mark Beattie's 'Double Helix Sculpture', for instance, which was shown at the Serpentine gallery and won the Xavier prize in 2015, is reflective metal and is 2m tall.[38] Beattie's sculpture was made in response to a brief that asked for something 'architecturally inspired' and this link with architecture again connects with Jencks and the concept of the double helix being foundational to the structure of the body. The piece of sculpture which commemorates Crick in his birthplace of Northampton was a collaboration between Lucy Glenddining and the monument company m-tec. The sculpture is called 'Discovery' and features two intertwining steel pillars with human figures at the end of them.[39] It shares the aesthetic of optimism and expansion with Jencks and Beattie, and similarly an unendingness that all the artists incorporate. This sense that somehow DNA can combine past and future in its rendering is key to the way that it lives in the public imagination. These actions of commemoration are futuristic but respectful, serious and playful, optimistic and grave. The play between then and now in them is key, with the concept of DNA shimmering in a kind of historical interzone. Such artworks contrast for instance with the memorial to Maurice Wilkins found in his birthplace of Pongaroa in New Zealand, which is three stones on top of each other with a helix carved into each. This much more solid, static, organic version of the helix is uncommon (Figure 1.2).

FIGURE 1.2   Olafur Eliasson, *Umschreibung* (2004)

A development of the concept of simultaneity is found in Olafur Eliasson's *Umschreibung* (2004), a vast staircase that winds through a double helix before returning to its beginning, essentially a Mobius strip. The title means 'periphrasis' and the concept is about 'movement without destination'.[40] Similarly to Jencks, then, Eliasson presents DNA as active, participatory, fluid, and continuous. The work demands participation, too, ensuring that contemporary bodies move around the art. These public artworks add to the way that DNA is articulated in the public imaginary. Their status as public interventions similarly contributes to the wider monumentalisation of the double helix as part of commemoration culture. They testify to the dual nature of DNA in the public imaginary, as they are both commemorative and 'now'.

This aesthetic was developed during the undertaken by the design company SomeOne in 2015. On behalf of the Crick Institute (to help fundraise the new building), the company commissioned 20 sculptures of DNA structures, each decorated by a different international artist.[41] Some of the art was corporate – there is an Aston Martin and a Ted Baker one, for instance – but others were the work of important artists such as Zaha Hadid and Ai Weiwei, and the final pieces were auctioned at Christie's.[42] The art trail that they created through London encouraged a bodily experience of engaging with genetic data. At the same time the art trail raised consciousness of DNA in everyday settings and encouraged creative responses to genetic material. Simon Manchipp, Creative Director at SomeOne, pointed out the importance of the double helix as a design icon: 'When working to develop a series of sculptures to be customised by some of the world's leading creatives we considered many different kinds of form, but we kept on coming back to the DNA spiral'.[43] The double helix, then, here stands in for all of genetic science, becoming symbolic of genetic knowledge of all kinds; literally standing in for DNA in a world of signs. It becomes an icon of knowledge, something that has a status as a commemorative and symbolic act. The London helixes were slightly smaller scale than others (although still 7 feet tall) and decorated in numerous ways according to each artist. They create a public action of commemoration and bodily engagement with the symbolic representation of DNA.

The aesthetic of double helix public art described so far – out of scale, metal, gesturing to the ineffable, often including bodily movement – is replicated in the piece that marks the beginning of the DNA cycle path in Cambridge. Opened in 2005, this was collaboration between the Sanger Institute and Cambridgeshire County Council. In itself, this piece of public art (opened by Sir John Sulston) makes physical the link between body and DNA, celebrating the corporeal. The path is marked with 10257 stripes representing the nucleotides of the BRCA1 (Breast Cancer type 2 Susceptibility) gene. It is a way of rendering the size and scale of genetic information to a visitor-user. The participant cycles the route, re-enacting along the way the BRCA2 gene; it is two miles long. At each end, there are two sculptures of double helixes. The experience of cycling the entire route would be to somehow corporeally experience the gene, as well as linking the two sculpted helixes. The route is both material and bodily, the painted tarmac activated by the body of the cyclist. The route once again demonstrates how public commemoration and heritage culture has engaged with DNA, contributing to an evolving aesthetics of commemorative and public historical practice associated with genetics. Again the key thing here is the interaction of the body with an upscaled representation of the data. The route also links to Francis Crick Avenue and on to the Addenbrooke's hospital complex, combining geospatial location with commemoration and contemporary scientific research.

This route features in Ali Smith's book *How to be both*, discussed in Chapter 5, and the novel's discussion of the interrelationship between imagined and real in the public commemoration of DNA is useful in considering the ways that genetics figure in the public imagination. The protagonist George cycles out to see a structure she's noticed from the train: 'It *was* a DNA structure after all, a sculpture of one, and it marked the start of a cycle trail you could follow for two miles along the little different-coloured rectangles painted on the tarmac, each standing for one of the 10,257 components there are in a single gene' (p. 171). George cycles the length of the gene, taking a video on her phone and making a participatory artwork of it. Her interaction with the artistic DNA is digital and physical simultaneously. The sculpture makes her think about history and representation:

> It resembled a joyful bedspring or a bespoke ladder. It was like a kind of shout, if a shout to the sky could be said to look like something. It looked like the opposite of history, though they were always going on at school about how DNA history had been made here in this city.
>
> What if history, instead, *was* that shout, that upward spring, that staircase-ladder thing, and everybody was just used to calling something quite different the word history? What if received notions of history were deceptive?
>
> Deceived notions. Ha.
>
> Maybe anything that forced or pushed such a spring back down or blocked the upward shout of it was opposed to the making of what history really was. (p. 172).

What is referred to here as 'DNA history' is of the old-fashioned, technical type – the material actuality of the work of Watson, and Crick at the Cavendish Laboratory in Cambridge rather than the process of it or the interpretation. This is the type of history taught in school, pedagogic and dogmatic and dry ('they were always going on at school'). It consists of events, people, and definable objects, is measurable and teachable in digestible chunks. Such practice of memory is very unlike the cycle path, which simultaneously engages time, linearity, body, and mind to celebrate corporeally and remind the participant of their own place in temporality.

In resistance to this dry 'history' is a strange, indefinable form. It 'resembled' other things, cannot be understood wholly, its meaning is deferred through comparison. It is a shout, something purposeful but without seeming form, strange and challenging. It is a 'joyful bedspring', something inanimate given emotion. This 'joyful bedspring' identifies the strange quality of DNA and the double helix structure. It is useful, inanimate, odd, out of place, absurd, and basic. Without it the structure would collapse, but it has a jaunty extra aura somehow. This sense of playfulness is key to the way that the book conceives of DNA. It is structural, fundamental, but slightly cheeky and potentially absurd. It challenges the world to not find it uplifting, literally in the sense of the 'bespoke ladder'.

This strange sculpture stands as a challenge to the dry history of information. Smith's protagonist suggests that data, form, materiality might be suggested as the 'opposite of history'. George has already found 'history' upsetting: 'George is appalled by history, its only redeeming feature being that it tends to be well and truly over' (p. 104). For her history is 'a mound of bodies pressing down into the ground below cities and towns in unending wars' (pp. 103-4). In contrast, the jaunty structure suggests something else. Life is the challenge, the spring, the shout, and 'history' drags it down, attempts to render it understandable and measurable. 'Received' history attempts to distort the shout, to push the spring back. DNA has the potential to queer 'received notions', to demonstrate the deception in the way that people have been taught to think (or have come to think) about the past. In contrast, it is strange, uncanny, odd, something in-describable, weird, fun, physical, and imagined. Here, the body at its most fun-damental, that is, the 'building blocks' of life, challenges the epistemological rational, goading sense, and attacking realism. In place of 'history', we get the absurdity of a 'joyful bedspring', the weirdness of the double helix structure, and the bodily experience of cycling the gene.

The Thing Itself

The importance and spectrality of the physical to imagining genetics is demon-strated by the exhibit of the DNA structure in the Science Museum (London). The main object relating to DNA in the central 'Making The Modern World' gallery is the DNA molecular model made by Crick and Watson in 1953 whilst working through their theories.[44] The gallery is a space devoted to 'iconic objects' that have 'helped to shape the world we see today'.[45] In itself, this curatorial desire to present relevant icons demonstrates something about the way that science heritage communication works in museums. The model is trailed on the Science Museum website as being a key part of the gallery: 'Chart 250 years of science and technology [...] Come face-to-face with the Apollo 10 command module, Babbage's Difference Engine No. 1, Crick and Watson's DNA model and the first Apple computer'.[46] The emphasis is on the encounter, the physical engagement 'face-to-face' with the important object itself. Key here is a discourse of au-thenticity and significance within the heritage space. This is *the* model, rather than a representation. It is important for its scientific significance and also for the historical link it allows from now to then. The object here attains an aura of importance despite – particularly in the DNA model's case – being sometimes quite visually unimpressive.

Crick and Watson's model is a flimsy, insubstantial thing, easily passed by and missed. The model is overshadowed by most of the objects around it. It is see-through and has little evident purpose. The model itself is made of aluminium alloy with annotations drawn onto plates. The historical significance is, of course, that the model itself is part of the proof that Crick and Watson provided in their work on DNA. The three-dimensional structure of DNA was theorised and their physical model demonstrated how it worked. Crick wrote in his memoir that 'Following Pauling's example, we believed the way to solve the structure was to build models'.[47] Indeed Crick points out that Watson happened upon the correct model 'not by logic but by serendipity', something which 'demonstrates that play is often important in research' (pp. 65, 66). Crick's testimony is that this was nearly accidental so the significance of the thing itself is slightly undermined. However, this 'model' is the thing itself, the actual object that Crick and Watson made in 1953 and which led to their understanding of DNA's dimensional structure. The insubstantiality belies its significance, so as an educational exhibit the piece is highly important but as an 'icon' it lacks punch.

The model is a representation of something that (at the time) was not verifiable visually and which was indeed only being theorised. It is a way of 'creating' and 'writing' the theory of DNA, a speculation. Modelling is a way of structuring knowledge, of imposing a version of events and articulating, physically, the thing that is being discussed. It is a way of understanding and envisaging structure. Modelling is also a way of knowing and of imposing epistemological order. Here, the model itself structures knowledge, capturing in its dimensionality a means of knowing and articulating the world. The DNA structure, in all its insubstantiality, stands in for the thing itself, and eventually comes to 'represent' this knowledge (as can be seen in public art, discussed elsewhere in this chapter). This is a case where 'structure' is coexistent with 'knowledge', as without the understanding of the structure of DNA Crick and Watson would not have had knowledge of it. The two are constitutive of each other and in the DNA model at the Science Museum we see the then and the now of DNA, the movement of this thing conceptually and imaginatively between physical and ethereal, from then and now, from theory to practice. Rita Felski argues, 'Cross-temporal networks mess up the tidiness of our periodising schemes, forcing us to acknowledge affinity and proximity alongside difference, to grapple with the coevalness and connectedness of past and present'.[48] The insubstantiality of the DNA model – you look straight through it – raises this concern, the model that is seemingly unfinished, the structure that is being described, the significance of the thing which appears to be purposeless. This model is not DNA, but it does 'represent' DNA, or give us a language of DNA that can be utilised and understood. It is not DNA, but it does give a re-presentational nexus point so that everyone post-1953 can conceptualise, visualise, and understand how their lives are structured by this molecule. The seemingly contradictory or circular point I am struggling to make is that DNA is given form by this model which is itself is of the form of DNA which itself only has form in our minds because it has been modelled. The model is the proof and the model is the museum exhibit, and both coexist within a complex discourse of iteration.

Crucially, the model is incomplete. This is because the 'original' was taken apart shortly after their work to be repurposed. The object in the museum is itself a reconstruction, partially recreated (after the 'original' plates were discovered by chance). For an icon, then, this piece is quite insubstantial. It is a re-enactment of something that was significant in 1953, built from the original materials but not actually the 'thing' itself as it was 'originally' created but a later construction. So the museum's model of DNA is not actually 'authentic', although it does have the aura of being the 'original' pieces. This model which allows us to conceptualise DNA is itself replicated, a literally empty signifier which has been put back together again to signify within the museum heritage context and to communicate therefore something educational. The model addresses the concept of authenticity and representation whilst itself not being 'real' despite being exhibited due to its material significance as something important. It is therefore highly contradictory and in this communicates something of the ethereality and contradiction of re-presenting and remembering DNA.

A further representational point is that this is the model that appears in several photographs of Crick and Watson taken by Anthony Barrington Brown. These photographs were taken in the Cavendish Laboratory in May 1953 just after the publication of their *Nature* article.[49] The images, intended for *Time* magazine, have become themselves iconic after they were published in Watson's bestselling 1968 memoir *The Double Helix*.[50] Barrington Brown's images contribute to the iconisation of DNA and in particular the association of Crick and Watson with the heroic male 'race' to describe – to model – the structure of the molecule. It has been suggested that the images were used to reconstruct the model when it was put up in the Science Museum although whether this is true is not clear. This notwithstanding the rather sad model in the gallery is considered as an 'icon' by the Science Museum and contributes to the understanding of DNA as something that has become post-representational and without signification. It is important to recognise that the popular image of DNA has not changed greatly since 1953, or Watson's memoir of 1968. The double helix has become an icon, a signifier without signified, coming to mean without reference to actuality. This matters as the untethering of the thing from its symbol leads to a wider misunderstanding. As DNA becomes something historicised, then rather than now, a heritage object worthy of commemoration, it loses substantiality in the contemporary, referring always and only to its moment of original conception and modelling. It becomes something to be imagined and represented. This example shows us how the manifestation of genetic materials and concepts can illuminate discussion of public history and commemoration culture. In particular, this exhibit shows the continuing significance of the 'original' in understanding historical events. It also shows the continuing resonance of models and three-dimensional representations to communicating scientific breakthrough, and comments therefore on the ways in which the double helix has become so representationally important. The exhibit shows us the ways in which the double helix began to become iconic in representation. Untangling the iconography here also allows us to understand how the structure of DNA is bound up with its historical representation.

Revising Rosalind Franklin

A clear change in the way that DNA is remembered and historically 'placed' has happened over the past two decades in relation to the reputation of Rosalind Franklin. Franklin's status and association with the 'discovery' of DNA has changed and developed, and she is now relatively well known where before she had been marginalised. As part of a wider reclaiming of marginalised female scientists and a championing of women's work, Franklin has been recognised as key to the recognition of the double helix structure. This is due as well as an increasing understanding of scientific process as collaborative effort. This is the irony of Franklin's position. She is both a feminist icon and prime mover, and an enabler and scientific heroine. Her status as a corrective to the normative storyline of male achievement, means that her example contributes greatly to the historical imaginary and hence inflects the way that DNA is conceptualised and considered as an historical entity. Understanding the case of Franklin enables us to see how genetic history is participant in reclaiming memory. Franklin's example gives us insight into how to remember and recollect DNA. Franklin's example shows short-term revisionism in restitution, and the ways that public historical practices can be deployed instead of real change. Expressions of regret around how she was treated and reclamation of her name and legacy are not linked with shifts of policy or practice (Figure 1.3).

Franklin was the first person to 'see' DNA and her work hastened it into 'being'. Before her work everything was theoretical; she provided the data that enabled Crick and Watson to build their model of the double helix (discussed below). In 1953 images taken by Franklin's student Ray Gosling using her advanced x-ray crystallography techniques were shown to Francis Crick without her knowledge. Crick and Watson also were given some of Franklin's data, again without her knowing although the circumstances are contested. Franklin was studying DNA from one particular point of view; because of their own particular model-led approach, the images and the data in part led to Crick and Watson's understanding of the double helix structure. The seemingly key image has since become iconic, known as 'Photograph 51', and indeed it often stands in for all of Franklin's work despite her only working on DNA for two years and having more important success in other areas (it is the image that was used on the commemorative 50p piece minted in the UK in 2020). Watson and Crick's 1953 article clearly refers to her: 'We have been stimulated by a knowledge of the general nature of the unpublished experimental results and ideas of Dr. M. H. F. Wilkins, Dr. R. E. Franklin and their co-workers at King's College', London'.[51] This seeming marginalisation is complicated by the fact that Franklin published a paper in the same 1953 issue of *Nature*.[52] She added a note to the submitted

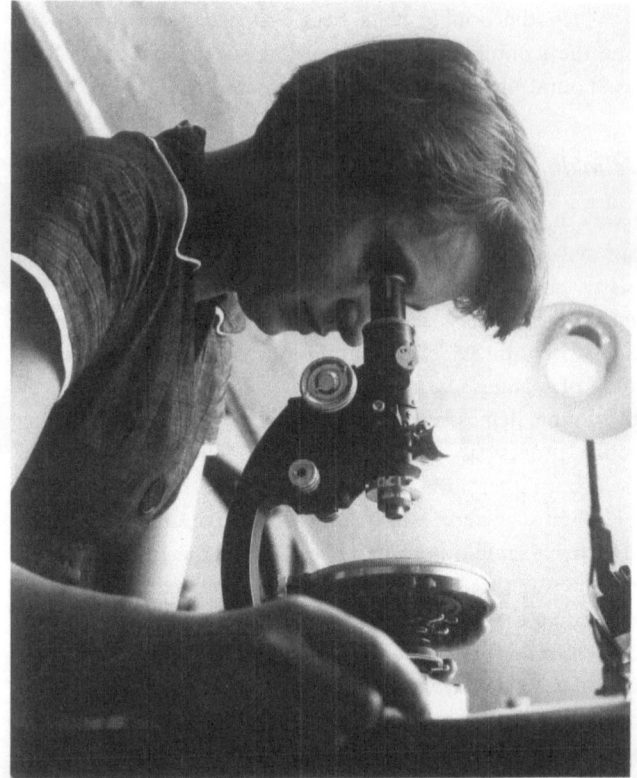

FIGURE 1.3 Rosalind Franklin with a Microscope in 1955

Creative Commons Attribution-Share Alike 4.0 International license, https://commons.wikimedia.org/wiki/File:Rosalind_Franklin.jpg.

Credit: From the personal collection of Jenifer Glynn/MRC Laboratory of Molecular Biology.

manuscript when she heard about the double helix announcement suggesting that 'Thus our general ideas are not inconsistent with the model proposed by Watson and Crick in the preceding communication' (p. 741), and indeed as her data had underpinned their argument so her article evidently supported such assertions. As Brenda Lennox argues, 'there is no evidence that Franklin felt bitter about their achievement or had any sense of having been outrun in a race that nobody but Watson and Crick knew was a race'.[53] Part of the problem of distinguishing Franklin's achievements is that the model of heroic success is simply not something that was important to the working practices at the time, or subsequently, in scientific research. DNA was not even seen as particularly important in the early 1950s; Franklin soon moved to work on the structure of viruses, work which is enduringly important.

Yet it is the case that Franklin's foundational work on DNA has often been seen to be occluded or ignored in accounts of the events, and that the memory of the double helix discovery has become part of a key narrative of scientific achievement and progress. Franklin died in 1958 and was to a certain extent forgotten and marginalised. Crick, Watson and Wilkins were given the Nobel prize in 1962 but Franklin couldn't be nominated as posthumous nominations are not allowed. Subsequently Franklin's reputation has been argued over and her role discussed at length. Partly this is because of the way that Watson wrote about Franklin in his 1968 memoir. Watson admitted that 'Rosy, of course, did not directly give us her data. For that matter, no one at King's realised they were in our hands'.[54] Watson's patronising tone, condescension ('Rosy, as we called her at a distance'), characterisation of Franklin as Maurice Wilkins's assistant and as stiff and unyielding, and of course his free admission that he and Crick had used her data without her knowledge all contributed to a public sense that Franklin had been hard done by. Watson's book led directly to the 1975 publication of *Rosalind Franklin and DNA* by her friend Anne Sayre, a reclamation of Franklin as a serious figure in her own right.[55] When Watson reprinted his book in 1996 he added an 'Epilogue' which included tribute to Franklin's courage, intellect, and expertise. Watson admits that his 'impressions' of Franklin were 'often wrong', arguing that Crick and he understood 'too late the struggles that the intelligent woman faces to be accepted by a scientific world' (p. 225) (Figure 1.4).

Franklin is a complex figure to approach when considering the public understanding of the history of DNA. She is a revisionist figure, insofar as she suggests the need for a retelling of what seems to be a familiar story, and a recapitulation of what is important in public memory. As the renaming of streets in Norwich and halls in Portsmouth show, Franklin is now used as a clear alternative to Watson. Her name signifies a revision to public memory, an effacing of a problematic figure with one considered more suited to contemporary narratives. Franklin is being used here to change public memory of scientific achievement, with contemporary commemorative practice choosing to highlight her rather than that of her male peers. She stands as a criticism of the male, positivist, teleological version of science history, her status in public discussion as a corrective to this. On the one hand, there has been a blue plaque outside her former home since 1992 commemorating her contribution to 'the study of molecular structures including DNA'. On the other hand, the plaque on the Eagle pub in Cambridge, which commemorates that Crick and Watson announced the double helix there has been vandalised in 2017 to add '+ Franklin' (a plaque had been added inside the pub outlining her contribution in 2014).[56] The Institute that bears her name and King's College, where she did some of her most famous work (but left as soon as she could to join Birkbeck) hosted a series of events celebrating her centenary year in 2020. She was the subject of a Google Doodle on her 93rd birthday, an image that was used around the world. She is increasingly becoming conceived of as an important figure in her own right, with a public profile and a contribution to the historical imagination. In particular, the image of 'Photograph 51' is increasingly used and is now clearly part of the understanding of the historical development of our knowledge of DNA.

FIGURE 1.4   Controversy about Franklin's Memory

Franklin's reputation is now higher than ever but often tinged with an imposed melancholy. When the European Space Agency ExoMars Rover was named 'Rosalind Franklin' in 2019 the two aspects of her public memory were referred to. On the one hand, the UK government celebrated her as 'UK scientist and co-discoverer of the structure of DNA'.[57] On the other, in the press release announcing the new name the UK Science Minister Chris Skidmore closely linked Franklin's presumed marginalisation to the narrative of the naming: 'Just as Rosalind Franklin overcame many obstacles during her career, I hope 'Rosalind the rover' will successfully persevere in this exciting adventure, inspiring generations of female scientists and engineers to come'. Franklin's public memory, then, is changing (she is now the 'co-discoverer' of the double helix) but linked with this rehabilitation is a clear sense of her presumed marginalisation. During discussion of whether she should be on the £50 note her experience was regularly presented as a warning: 'Franklin's story is now seen as a cautionary tale about the ways the contributions of women have been ignored over the course of scientific history'.[58] The same story quoted Katherine Mathieson, CEO of the British Science Association, arguing that 'we could set the record straight' and understand Franklin's achievements. The public discussion over the note demonstrates the extent to which Franklin's public reputation is imbricated with her having 'lost out' in the 'race' for DNA. Yet some of the public discussion cited her as simply key to the discovery of the double helix, with no reference to the controversy over the use of her data. Franklin's reputation is changing and developing, and hence the ways in which genetic science, its key moments, and its importance, is shifting too. In this example then we see DNA as part of public history, contributing to an ongoing discussion about how and why a society remembers. From a public history perspective the reclamation and development of Franklin's reputation is an important act, and it allows us to see how cultures and societies conceive of the past, and the ways in which genetic science both can contribute to this and also inflect it.

This complexity of the historical genetic imaginary can be clearly seen in the case of Anna Ziegler's play *Photograph 51*. The play, which was developed in 2008 and premiered in New York in 2010, strives to 'see' Franklin. Its revival in 2015 in London with Nicole Kidman as Franklin was an important moment in the public and popular rendering of her life and reputation. As has been seen, Franklin's public reputation is a complex thing. Critics responding to the play mentioned how Franklin's 'vital contribution to uncovering the structure of DNA has been marginalised', her work 'had been, and still are to a great extent, sadly overlooked', and how 'she has never been given the credit accorded to her rivals and colleagues'.[59] Indeed, Steve Connor writes that Ziegler's play 'attempts to redress the balance' and hence sees the text as a revisionist drama. It tells us something occluded, unseen, important but somehow unperceived, and reclaims Franklin from obscurity. In these critical constructions Franklin to a certain extent disappears into this reputation as the forgotten woman, defined still by what she did not do (win the Nobel Prize) rather than what she did, the victim of injustice rather than an excellent scientist. At the same time the work to re-remember Franklin is an important recovery.

Photograph 51 makes DNA into a mysterious thing again, something pre-double helix, something mystifying that cannot be reduced to a simple model; understanding it is the stuff of a life's work. It is part of the re-historicising of DNA, a way of recollecting that such scientific work was produced at a very precise historical moment. Ziegler focuses on the 'story of the race to the double helix'; however, she prefaces her play with a disclaimer: 'a work of fiction. I have altered timelines, facts and events, and recreated characters for dramatic purposes'.[60] It is a biographical drama on science, echoing Michael Frayn's *Copenhagen* (1998) in its complex rendering of theoretical science through dramatising human relationships. Revolving mainly around Franklin and Maurice Wilkins, and very fluidly staged, the play considers some key themes relating to DNA: what we can see, what we know, how we know it, how perspective matters, and the beauty of life. It also adds some particularly useful tropes from scientific biography: competition; the loneliness of research; the danger of pride.

The play resists the notion of DNA being 'authored' but instead shows how various characters, particularly Watson, self-dramatised the importance of particular moments. 'There's an element of fate to it' he says early in the play, 'To be born at the right time', p. 27; later, when Wilkins seems to reject working with him, he says 'Was it the biggest mistake of his life. Without question', p. 29. In particular, the moment Franklin sees Photograph 51 she 'becomes' different, somehow – although possibly without understanding it: 'And she stood there, staring at the photograph, as though she were looking in a mirror but was suddenly unrecognizable to herself' (p. 45). Drama often works by showing an audience the internal thoughts of a character – *Copenhagen*'s contribution is to suggest that no one knows anything about the world other than that which they perceive – and *Photograph 51* is in many ways about not understanding what one can see, the distinction between what is visual and what is known. In this it contributes to a wider discourse around how to remember, recollect, and imagine DNA. The ineffable aspect of seeing and not seeing something, the elusive quality of this 'thing' which is being described in the image, is inherent in the genetic imaginary.

The play also dramatises the urge to investigate, contrasting a male will to 'control' with Franklin's more stolid devotion to 'the work'. She is enthralled by the beauty of the natural world but not desiring to control it. In contrast, Don Caspar says this about X-ray photography: 'As though you and you alone possess the superhuman powers that allow you to see into the heart of things. To understand the nature of the world as though it's a secret no one else is meant to know' (p. 34). This sense of peering at nature is relatively common in scientific discourse, here given to a male character in echo of the hundreds of male scientists that have claimed to draw back the veil of nature. In contrast to this type of thinking about 'life', Ziegler suggests that male pomposity kept Franklin back, both in terms of preventing her from accessing key moments and also because her colleagues resented her manner and were less loyal to her than they might have been. She is at all times the only woman onstage, unique but also isolated and alone. Two of the male scientists seek to 'understand' and 'see' her properly, but they both fail eventually. She resists this and is unknown to them – but not the audience, who 'see' her real character in an unspoken moment late in the play when she acknowledges her loneliness to herself but does not say it out loud.

Franklin's loneliness, her solitariness, her unique, non-doubled quality, is emphasised throughout. Glibly the play argues that 'There's no science that can explain it. Loneliness' (p. 71), suggesting a kind of melancholy that can't be understood – 'seen' – by scientific means. By making the search for DNA something that is expressed through one woman's lonely desire for meaning, the play personalises the science, makes it emotionally comprehendible, something that the audience can 'see'. The human is the model of the DNA, the emotional body that navigates between molecule and time. Franklin's last, sad speech before her death – in answer to the question 'what do you want?' – reveals the play's concern with what has been lost, rather than what might be found: 'to be kissed, to feel important, to learn how to be okay being with other people, and also how to be alone. To be a child again, held up and admired, the world full of endless future' (p. 71). Throughout Franklin and Wilkins discuss Peter Brook's 1951 production of *The Winter's Tale* and this is an important intertext about sadness, loss, misplaced hope (although the joke – a meta-joke about Franklin and her status as the lost female scientist – is that neither of them can remember who played Hermione in the production (it was Diana Wynyard). The play seems to suggest that Franklin was unfulfilled and that her work was simply not enough – or that she worked so hard due to an unacknowledged emptiness. However, the sense of enquiry being a way of giving shape and meaning to that which is unknowable – life – is a key motif to the work. As she says before she dies, "We lose. In the end, we lose. The work is never finished' (p. 75).

Photograph 51 contributes to the development of Franklin's public reputation and the various ways she might be imagined and celebrated. The play's concern with what is seen, and unseen, and with how different ways of looking might reveal many different things, similarly contributes to the wider comprehension of how visualisation contributed to the recognition of DNA ('to Rosalind, making a model was tantamount to negligence' says Crick, p. 38). The play's fluid range of concepts – it reaches post-war Britain, ambition, gender, investigation, plagiarism, sadness, and higher mathematics – means that it is self-consciously complex. Historiographically it suggests that which has not been seen, and also reminds us that 'facts', even in relation to the 'revealing' of something true and real, might be more complicated than they first seem. However it also suggests that discovery is something that can only be understood and explained through a range of considerations – personal, chemical, economic, social, gendered. This then is the contribution to public history that DNA might make, rather than simply the public history of DNA. The contrast between what is known and what is seen, the complicated reveal, suggest to a viewer the insubstantiality of our understanding of the past. Even in the glare of the laboratory is it not completely clear what has happened, why, to whom, and at what cost. Franklin's reputation is complex and demonstrates the ways that 'public' history is made and articulated according to particular biases. *Photograph 51* suggests that the history of the double helix is as much a history of the complicated interaction of human motivations as it is the account of scientific observation. The drama of DNA is found in the tension between model and thing, articulation and object. In the public historical imagination, it sits as something seen and unseen.

Seeing Richard III

The most famous Anglophone moment when genetic science and history came together as public practice was during the discovery, exhumation, and examination of King Richard III's remains in 2012.[61] The remains were found under the site of a car park in Leicester and identified through laborious DNA investigation. The identification of site and the remains involved a great deal of archaeological science. The investigation into Richard III's remains after they were found was a complex project intellectually, scientifically, and ethically.[62] The work on the DNA of the skeleton was crucial in establishing the veracity of the bones. This highly public work prompted a set of popular discussions in the press and beyond about the nature of lineage, the contemporary world's relationship with heritage, and the ethics (and legality) of disinterring human remains. As has been pointed out by critics, much of the imaginative work has been to rehabilitate Richard from the 'received' collective memory. The bones and the genetic analysis are made to do a lot of work in terms of interpretation. Richard emerges anew from this work, now a worthy King and a tourist attraction. The development of a museum devoted to Richard III and his further linked public historical practice with genetic investigation, as a commemorative and heritage site was developed to narrate a type of double helix history, as this section will outline.

Discovery and Analysis

King Richard III died in battle in 1485 during the last major battle of the Wars of the Roses. His body was taken from Bosworth field and buried in Greyfriars Friary in Leicester. The burial site was built over and the precise location lost. By the 1940s the site had become municipal offices and an employee car park. Richard is a notorious figure in the English national memory. Allegedly a hunchback, successive characterisations of him as a villain – notably in Shakespeare's *Richard III* (1593) and a variety of film adaptations – have led to a remarkably resilient collective memory. This reputation is also due to the disappearance of his nephews whilst he was Lord Protector. The two are known as the 'Princes in the Tower' and it has been long assumed that Richard murdered them in order that he could maintain his grip on the throne. Members of the Richard III society had claimed for some time that the King's grave site was under the car park in Leicester. The local history society was formed to rehabilitate the king's reputation in response to what they called 'propaganda'.[63] Over the period 2004–11 the society began to seriously argue that the site was correct and should be investigated; their hope was that 'new attention drawn to Richard by the discovery will inspire a reappraisal that could rehabilitate the medieval king'.[64] In partnership with various organisations, including the University of Leicester and the Leicester City Council fundraising, began to undertake the project, with excavation starting in 2012.[65] This shows the importance of local history societies in challenging mainstream views and bringing their particular focus to investigations. There was considerable global media attention paid to the project.

On the first day of digging the remains of a male body were discovered. These remains were subject to various types of analysis, as will be outlined below. Most significantly, a small amount of DNA was retrieved and looked at by researchers in Leicester led by Turi King. Extensive work demonstrated that the DNA matched that of Richard III's maternal line, meaning that the evidence suggested the remains were that of the king. News reports around the world used the DNA evidence as confirmation that the skeleton was that of the dead king.[66] The bones took on a kind of material substance as they were proven to be those of Richard III, shifting their physical nature into something to be celebrated and dreamed of, something that had relation to the contemporary world.[67] The discovery of the King's remains suggested that genetics might allow the identification of 'truth' in the past, particularly in relation to physical identity. Most importantly for our analysis the majority of the press associated the historical investigation and associated revisionist account of Richard with the insights provided by genetic science. Without the use of DNA analysis as an historical tool which proved 'beyond reasonable doubt' that the remains were Richard, the detective work would not have provided a clear resolution.

DNA Discovery Narrated: The King in the Car Park

Channel 4 had cameras on the spot when the remains of the body were found. They had been covering the project having been contacted by some of the key participants. This meant that they were first on the scene and also had access to the whole identification process. The resulting documentary, *The King in the Car Park*, was screened in 2013.[68] The film follows the extraordinary events of the disinterring of the body of Richard III. Furthermore, it opens up some of the key issues around this work. The audience is made to recognise the historical work that is being done in turning a set of muddy bones into an internationally celebrated King, following the journey as the DNA investigation translates buried remains into an historically significant identity.

The tone of the documentary is uneven, communicating a general unease about how to approach this topic. On the one hand, it is serious history and worthy of some study. On the other, the celebrity link and the excess of the story make it extraordinary and strange. The film is fronted by Simon Farnaby, an actor-comedian who self-confessedly does not know much about Richard III. His consciously amateur approach figures the presenter as everyman investigator, sharing the audience's innocence: 'But the more you look into the historical side of things [...] the murkier and murkier it gets [...] but that's what I've learnt to love about history'. This tonal combination was noticed by critics who argued that this was something particularly right: 'if the tone of the programme felt confused [...] it ultimately all felt rather apt, even charming'.[69] Reviewers presented the documentary as of a piece with the whole story – that is, communicating something peculiarly English. The investigation and commemoration of this body, this king, communicates an eccentric national character. This is the value of the documentary, as it presents the sheer oddness of the spectacle as it unfolded. It is both banal and extraordinary, as the title of the film communicates. The bathetic mix of day to day and regal suggests something about contemporary historical experience in England and how genetic analysis has contributed to it.

The original investigation into the resting place was initially taken up by the Richard III society. Once the area had been identified they focussed on ensuring it was investigated. This amateur society raised the money for the dig in two weeks via crowdsourcing, demonstrating the abiding interest in this King and also the continuing importance of amateur historical organisations for driving change. The key figure in this is Philippa Langley, an obsessive figure who drives the dig by sheer force of will. Langley's interventions introduce an affective, non-rational aspect to the investigation: 'I got the strangest sensation when I was in that area, in that place, I absolutely knew that I was standing on Richard's grave'. She cries when the body is brought up, covers the bones with an approximation of the Plantagenet standard, and claims 'He was ready to be found, he wanted to be found'. The documentary does not quite know what to do with Langley. She is evidently right, and impassioned, and an important voice. She is also presented as eccentric and preferring her emotions to rational inquiry. Farnaby's voiceover talks about 'this seemingly bonkers project might just have pulled off the impossible' and the slangy suggestion that the whole investigation is a bit 'bizarre', coupled with amazement that it is right, communicates the tonal uncertainty. This strange quality is part of the approach to the body and to the figure of Richard. The bones *mean* in multiple ways, and the film communicates an indeterminacy and a discomfort in approaching such materials in historical documentary. The film also communicates the uncanniness of such archaeology and genetic investigation.

In order not to contaminate DNA the diggers use cleansuits. This dehumanises those digging and echoes the aesthetic of crime scene television. This is particularly communicated with the focus on the remains. The purpose and point of the documentary is this set of bones, and they variously are abject, mud, presented, investigated, put back together, highlighted, and identified. They are the mysterious centre of focus, made to signify and mean. Initially the bones, the remains, are clearly outlined and looked at. The camera surveys them and has height on the abject materials in the trench. The aged material is turned into sample, created as something clinical through the clothing of the archaeologists. The contrast between the ways that the professionals work and the 'human', communicated by Farnaby, is both jokey and also a serious interrogation of the ways in which the body is turned into a set of samples. The camera spends a lot of time looking into the trench, resting on the muddy bones. Farnaby suggests 'I suppose it's a weird thing digging up a dead body' but adds 'It might not be Richard but whoever it is … was a person'. The ethics of the documentary and the disinterring the body, the work on the samples, are highlighted here. The audience is forced to recognise the dirty work of history. Farnaby is moved, as the camera peers into the trench:

> Its quite emotional thing isn't it really, it suddenly struck me just now coming and peering in, this was someone's relative, husband, father, brother, son.

This shift between thing to person, from object to human, articulates the tension in televising archaeology and in particular televising genetics. The abject materials were once a human but understanding the link between them and the living is difficult to see. The thing that links them is storytelling, narrative, historying as the signifiers are linked to signification and made to mean. Hence the film shows the ways in which evidence is made into meaning, given life, interrogated and interpreted. Furthermore, Farnaby's response to the bones shows the intrusion of emotion into historical investigation. In contrast with the archaeologists in their clean suits, he has a response to the bones which is affective. The bones are not samples but human remains, and the relative glibness with which the programme had previously dealt with them comes under some scrutiny. Why not simply leave the dead where they are? Again, the show's bathos is important here. Richard is a king, with a glorious and important legacy and memory. He is also a set of decayed bones in a box in the boot of a car.

The skeleton is dug around, marked, identified, and eventually removed from its resting place. The DNA test is discussed, particularly whether it is the right individual: 'it just feels a bit of an inappropriate thing to have done'. This renders the investigation strange. If it is Richard III, then this invasive work is fine; if not, then it is inappropriate. At this point the show is also reminding the audience of the massive press interest in the trench and the body, suggesting the pressure to identify, the importance of the 'right' historical answer, and the possible transformative effect of the genetic investigation at this point. Nothing else can give the world the 'correct' answer.

The film's middle section includes work on how the memory of the king developed after his death, including 'the Shakespearean myth'. It therefore uncovers the way in which reputation is created, and in particular how history is made and maintained. The programme thus participates in popular historiography, insofar as it reflects upon the way in which history is constructed in public. It also presents a revisionist account of Richard, as much of the investigation did. By looking at the physical, the material evidence, much of what has been thought or argued since his death can be reconsidered. DNA then is a contributor to this longer running discussion about reputation, history-making, nationhood, and popular memory.

Once in the laboratory the skeleton is worked on in various ways. It is situated on a lightbox, stark against the white light. The camera lingers on it repeatedly, as, again, we are reminded that this is the focus of study, the thing itself, the point of it all, the history in the show. There are many scientific suggestions about the bones, the skeleton, including the fact that parts of the skeleton appear feminine – as gender becomes something interpretable ossuarily. The bones are drilled into for carbon dating, clinically treated in various ways, to the extent that Langley walks out, saying 'I don't see bones on that table I see a man, I see a living, breathing human being'. Again the laboratory work contrasts the dead objects with precise equipment and techniques, ways of 'reading' the skeleton.

In order to 'prove' that the bones were Richard III's a living descendant needed to be found. This is key to the *truth* of the investigation: 'One final piece of proof was needed. If his ancient DNA didn't match a descendant then his identity would always be questioned'. Genetic evidence here is conceived of as the final contribution, closing off discussion. It also links the contemporary to the ancient.

They contact the descendant and he takes the test: 'If DNA from our bones matched Michael's, our skeleton would be a king'. The dramatising of the genetic testing process is shown, as the descendant swabs his mouth. The audience is reminded of the smallness of the action, of the way that genetics are in saliva. This man, seventeen generations apart from the historical Richard III, articulates the strangeness of genetic relationality: 'part of my physical being is directly related [...] it makes you stop and think, and it is somewhat overwhelming at times'. The investigation into his genetic data provides a match: 'The DNA evidence points to this being RIII'. The DNA is used to create a three-dimensional facial image of the King, demonstrating the ongoing need for a visualisation.

The documentary certainly does not present genetic evidence as the only tool involved here. Indeed, it is the embedding of genetic work as part of a suite of scientific approaches that is key. Genetics is the tool that can bring a kind of proof to historical investigation. However, it is also part of a wider discourse of understanding the body in the past through scientific technique. Finally, genetics is clearly here part of historical investigation, strange but normalised, here to stay as a way of gathering evidence about and interpreting the past. The documentary celebrates the novelty and impact of genetics in this context. The film also suggests that this is now the mainstream, that the argument about whether to use DNA in such investigations is moot. At the conclusion the combination of much evidence – trauma, carbon dating, forensics, skeleton – leads the investigators to 'suggest very strongly that this is him'. As Farnaby concludes, there are two types of investigation going on here, 'the science side and the historical side'. The science 'gets more concrete' but both contribute to meaning and help 'get the truth out'. Whilst they may contribute to the final 'truth', history and science are approaches that are divided here. The suite of approaches that read the skeleton of the King are situated within recognisably scientific spaces rather than the tropes most associated with history in documentary (library, maps, historical space) are used, but the skeleton is not seen there. Thus, there is something of a division between the 'clinical' and bodily – understood by science – and the reputational or narrative, understood and presented by historians. Genetic ways of reading these bones contribute to this scientifising of history. Genetics here understands the human body in time and space in a different way to that of the historian or even the archaeologist. It provides different evidence, and, importantly, asks different questions. The 'type' of historical knowledge that genetic science allows is useful for understanding of the past, and it provides new insight. The documentary demonstrates how the Richard III case accelerated the wider public understanding of the historical impact of genetic research. DNA evidence not only takes its place as part of the ways that the past is known, it also generates new and revelatory knowledge about that past. It is a tool to know the past and a means of presenting that past, of engaging with it. DNA here turns an abject set of bones into a King worthy of pomp and circumstance. It renders historical questions answered, and brings a rational discourse relating to proof and certainty to the debate.

Richard Reinterred and Rehabilitated

The body of Richard III was re-interred in Leicester Cathedral after a year of legal argument about the final burial site, including a final judicial review and decision made by the High Court. The legal challenge was made on behalf of the 'Plantagenet Alliance', a group who claimed to be descendants of Richard III and argued that he should be reinterred in York. The legal discussion illustrates the significance of the site of his burial, and the considerable public profile about this case. It also demonstrates the complexity of 'ownership' of historical remains. Leicester city council used the opportunity to remodel and develop the area around the cathedral, moving a statue of Richard closer to the site and purchasing the freehold of the school opposite the grave in order that it might become a £4m visitor centre. Richard III became a key component of the city's tourism strategy. Tourism in Leicester increased significantly, with 150000 visitors to the initial temporary exhibition and a rise in tourism income of around £482 million in 2014.[70] The King Richard III Visitor Centre opened in 2014 on the basis of 80000 visitors a year.[71]

The five days of events celebrating the re-internment of the remains situated the event at the intersection of heritage performance, nationalistic memorial service, and historical re-enactment. The narrative of the reinterring became subsumed into heritage performance. The pageantry was conspicuous, including the coffin retracing a section of Richard's journey from Bosworth field to the site of his death. The procession towards the final rest was attended 35000 onlookers, as the coffin progressed through Leicester town centre.[72] A mass was said for the king by the Archbishop of Westminster; thousands of people visited his coffin whilst it lay in Leicester Cathedral.[73] The memorial service was attended by a number of celebrities and the Countess of Wessex. The order of service featured a message from the Queen, and the service was conducted by Justin Welby, the Archbishop of Canterbury.[74] 700 journalists covered the service from around the world. The Dean of Leicester cathedral told the BBC that 'I'll be very conscious about the weight of history and also creating a new story about Richard in Leicester'.[75] This was a major event in the rehabilitation of Richard's profile, a revisionist version of the king being presented. It was also an event which underlined a kind of monarchical nationalism, focusing on the celebration of an historically understood Englishness. This was a public performance of national identity, with concomitant themes of inheritance, legitimacy, and a sense of reclamation. These issues of historical understanding are inflected and made possible by the genetic investigation. Genetic historical work becomes intertwined with nationalist celebration and public history performance practice. The foundation for the pageantry is the genetic work, and the 'proof' that the evidence provides gives the celebration its backbone. The narrative of the reinterring became subsumed into heritage performance (Figure 1.5).

At the service the actor Benedict Cumberbatch read a poem written by the then poet laureate, Carol Ann Duffy (poem discussed in Chapter 5). Duffy argues

FIGURE 1.5 Tomb of Richard III in Leicester Cathedral

that Richard's bones create 'upon cold soil/ a human braille' (ll. 1-2), a material image that suggests there is meaning in the body. The idea of the skeleton being 'braille', a set of symbols which depend on touch and indentation, figures the return of the king as something beyond text, both material and absent. The phrase also suggests that the bones can be made to mean if read properly. Cumberbatch had just finished recording his performance of Shakespeare's *Richard III* in the BBC's *The Hollow Crown* (released 2016). It was also reported in the press at the time of the re-interring that Cumberbatch was related to the 'real' Richard III.[76] This confluence of enactments and echoes give Cumberbatch's performance of the poem an uncanny aspect, capturing perfectly the way that Richard 'lives' in the historical imagination as a combination of imagination and bodily, genetically definable material. He is at once real and represented, then and now. Further re-enactment and historically-informed performance was evident in the service – based on a fifteenth-century framework – and the placing of a fifteenth-century bible on the coffin. This shifting un-canniness is something that characterises the events around the discovery of the remains. Throughout the process the motif has that of telescoping between then and now, shifting from micro to macro. The effect is to continually remind the participant, reader, viewer, of the various layering effects of history and narra-tive. The shifting temporality reflects the fragmented continuums of genetic knowledge in the public imagination, ghostly and material, then and now, connective and proleptic.

The rehabilitation of Richard and the increase in his public profile led to a longer-term tourist opportunity. The Science Museum opened an exhibit entitled 'Richard III: Life, Death, and DNA' which featured the 3-D printout of the king's skeleton (the exhibit went on tour to Ireland). Most importantly, the Visitor Centre opened in Leicester in 2014. The museum makes much of the extra-ordinary circumstances of the archaeological find, concluding its tour in a room built over the grave site so that 'you can see for yourself the very spot where King Richard III remained undiscovered for so long'.[77] As such the museum emphasises the strangeness of Richard's burial and discovery. The museum emphasises the importance of place in the heritage experience. The focus is on the trench dug to reveal the skeleton, the means of understanding and interpreting the remains. It also is centred on an absence, given that the remains of the king are interred in Leicester Cathedral. The museum has a copy of the 3-D printed skeleton and the reconstructed head of Richard III on display. As with the documentary, discussed above, Richard's skeleton becomes something to be focussed on, a heritage item in a museum and the centre of the exhibition, rather than the remains of a human being.

The museum's narrative gives precedence to those experts – geneticists and archaeologists – who were involved in the find and the work on the skeleton. As such the centre is a museum of history as well as process, demonstrating in a heritage context how expertise can be applied to develop and affect historical narrative. Throughout the public history of Richard III's remains, whether in the museum or the Queen's memorial message, the role of the academic team is foregrounded. The museum makes the 'detective' work part of the exhibit, fo-cusing on the process of rediscovery as part of the story of Richard III. This is a kind of popular historiography, insofar as the museum renders explicit the ways in which 'evidence' is read and understood. Genetic information becomes part of this evidence, a means for 'reading' the past. However, at the same time the museum seeks 'truth', a teleologic movement towards the actuality of the remains being Richard. In this the genetic information contributes seeming hard fact to support textual and other research. The DNA work 'enables' the transformation of these bones into 'Richard III', and furthermore defends this designation from any challenge. In the historical process then the genetic investigation becomes articulative.

The case of Richard III shows a potent public construction of history. Richard III has been reclaimed, reconstructed, and re-enacted. Through the intercession of genetic science Richard has been translated and identified, and this has con-siderable historical impact, both in terms of what is known and in terms of collective memory. The events show the importance of local knowledge, and the ways that scientific investigation can be unknowingly harnessed towards pageant and heritage performance. Most importantly, the case highlights how genetic investigation can become twinned in the public imagination with historical investigation, and how it can challenge narratives and present new evidence.

Genetic Public History as Revision

In this final section, I consider the ways in which genetic science has changed and challenged the archive of memory. In particular, I extend the notion of 'revision' in public historical knowledge by showing how new data can shift and change received historical narratives.

Sally Hemings

Before the discovery of Richard III the most high-profile case of genetic investigation becoming entwined with historical understanding and revisionism was that of Sally Hemings and Thomas Jefferson. This arose from the work of Annette Gordon-Reed, whose 1997 book *Thomas Jefferson and Sally Hemings: an American Controversy* had sought to establish the truth about rumours that Jefferson had lived with his slave Hemings and had several children with her.[78] Gordon-Reed's book took seriously the stories handed down in the Hemings family, and her investigation gave credence to the fact that Hemings and Jefferson conducted a long-term relationship. The relationship had been suggested and discussed for decades but never proven. After the publication of the book a retired pathologist, Eugene Foster, worked on the DNA of the various descendants of Jefferson and Hemings.[79] The work was cautious, but suggested a genetic link between the two lines. It is now generally thought that Jefferson fathered all six of her children. This was a combination of historical work and DNA evidence, the latter having substantial power despite some challenges. Christine Kenneally suggests that the genetic work in 1998 on the Jefferson descendants was foundational in a type of revisionism: 'There was a sense that thanks to DNA, the actual lives of historical figures – or anyone else, for that matter – could no longer be hidden behind a wall of respectability and lies'.[80] The DNA evidence has contributed to a wider collective understanding that Jefferson and Hemings had a relationship for many years.[81]

The historical debate is about more than simply paternity. As Catherine Nash argues, the genetic evidence has been folded into what is a complex and problematic historical debate: 'Deeply racialized and sexual imaginations of purity, miscegenation, cross-racial union, illicit sex, and sexual power run through the long-standing controversy about this dispute' (p. 141). The arguments about Jefferson and his relationship with Hemings are part of wider debates about slavery, power, patriarchy, violence, and reparations, as well as considerations of legitimacy and legacy. In addition to that, Jefferson is a Founding Father, central to American self-conception. The challenges to his authority attempt to render the national identity of the USA more complex and to challenge the assumptions that such nationalism is predicated upon. Genetic information therefore challenges multiple normative structures: purity; collective memory; nationhood; patriarchy; race. DNA evidence can challenge long-held assumptions, fracture families, and undermine dominant discourse.

Ultimately, it can also help reorder historical priorities. Part of the consequence of the DNA work has been to open up the historical record to other voices. Lucia Stanton and Annette Gordon-Reed have both published books on Hemings and other African American families who lived at Jefferson's estate; Gordon-Reed won the Pulitzer prize for her 2008 book *The Hemingses of Monticello*.[82] Hemings herself has become the subject of further study and more extensive recollection.[83] Her rooms next to Jefferson's have been excavated at Monticello, and there are tours of the house which focus on her experience. A joint exhibition created by the Smithsonian Museum and Monticello, *Paradox of Liberty: Slavery at Jefferson's Monticello* has toured the USA since 2012.[84] Shannon Lanier (a descendant) and Jane Feldman collaborated on the children's book *Jefferson's Children*.[85] Revealing the story of Jefferson and Hemings's relationship has diversified American history and brought into the discussion female, black, slave experience.

Whilst much of this activity would have happened irrespective of the genetic information, the DNA work has given the case particular gravitas. The historical revision and shift are in due in part to the association of DNA with a type of unhistorical authority. Gordon-Reed wrote in the preface to a later edition of her book, despite so much documentary evidence 'I believed at the time (and now) that the bar to resolution was more psychological than evidentiary'.[86] She adds, 'As it turned out, science and history complement one another perfectly' ('Author's Note', paragraph 12). Her point about the psychological bar suggests something about the way that DNA evidence might intercede into historical debate. This type of link makes the case watertight, contributes a strong foundation. Gordon-Reed had suggested in the Preface to the first edition that the 'final word' would be reached through the 'miracles of modern science and the wonders of DNA research, and not because of any interpretation of documents and statements' (cited in 'Author's Note', paragraph 1). In this telling science is revelatory and spectacular, claiming an authority that mere historical texts can never achieve.

The African Burial Ground Project and Henry Louis Gates, Jr

In 1991 initial archaeological work was undertaken as preparation for the building of a new federal office in New York. Many intact buried bodies were discovered, and further work suggested that the location was the site of a burial ground for Africans that had been used from around 1690-1790. After several high-profile campaigns and court cases the site was designated a National Monument and was opened to visitors in 2007. Extensive bioarchaeological work has been undertaken on the remains recovered from the site, including DNA testing.[87] Other African burial sites in Charleston (Georgia) and Portsmouth (Massachusetts) have had similar DNA investigation undertaken. These investigations have expanded knowledge of the African diaspora in the United States and elsewhere. They highlighted the possibilities of using genetic information to expand our understanding of the diasporic experience and to account for the historical violence enacted upon these communities. The investigation of the past of the African diaspora using genetics was therefore one of the most high-profile aspects of DNA investigation of the past. In 2000 Rick Kittles began to offer tests to African Americans through his company AfricanAncestry.com.[88] The company still exists, with its marketing suggesting that it 'helps people of African descent recover their history, reconnect with their ancestors, and create a lasting legacy for future generations'.[89] This rhetoric of 'revelation' and future aspiration illustrates the key aspect of the work that is being done on the genetics of the African diaspora.

As a consequence of this work there was a growth in interest in African diasporic DNA and concomitant development of new understanding of previously unseen pasts enabled by genetic investigation. This was turned into a public historical intervention by the historian Henry Louis Gates, Jr, whose documentaries made politically revisionist genetic genealogy into high profile television. Since 2006, Louis Gates, Jr, has presented *African American Lives* (PBS, 2006-8) and *Finding Your Roots* (PBS, 2012-), popular shows in which genetic genealogy is used to investigate the backgrounds of celebrities. The shows investigate family history, much as *Who Do You Think You Are?* Across the world, but the key innovation is the use of genetic information. *African American Lives* specifically focused on reclaiming the stories of those whose ancestors had been effaced by 'official' history. His work specifically linked the use of DNA investigation to revisionist historical approach, filling the gap in evidence and enabling a new scholarship. In his book to accompany the series, *In Search of Our Roots*, Gates makes the telling point that 'standard' historiography has failed the African American community by ignoring them:

> their stories remain to be discovered, buried in dusty archives, seemingly unimportant to anyone but a distant family member [...] Until their stories are reconstructed and told, these ancestors of ours will not exist as human beings, as agents, as actors in the great drama that is American history because under slavery our ancestors had no names[90]

The genetic work, then, enables a way of resisting this anonymity, gainsaying such a powerful attempt to erase and forget. This enables a participation in wider national history. Furthermore, though, genetic data can link contemporary populations to a wider and deeper past, beyond what textual evidence might remain. Louis Gates, Jr, argues that 'the obliteration of a conscious knowledge of the African past in the daily life of the African in America was achieved almost totally' (paragraph 12). To remedy this, DNA can re-connect communities to their roots. This has an important historiographical consequence: 'Restoring the stories of the lives of the members of our extended families can directly transform the way that historians reassemble the larger narrative of the history of our people [...] ultimately changing the official narrative of American history itself' (paragraph 17). Furthermore, this redressing of what has been intentionally unseen is itself a political act: 'It also allows us a glimpse into that which has been stolen, hidden, or lost, in the collective history of the African American people' (paragraph 19).

Once again we see here the idea that knowledge lives within the contemporary if it was just 'seen' properly; once again we see the motif of DNA investigation as 'revelation' but also as bringing additional information and revising our historical perspective. For Louis Gates, Jr, DNA allows African Americans 'symbolically at least, to reverse the Middle Passage' (paragraph 14), providing history where there was none and obviating the lack of archival knowledge. Indeed genetic investigation points out the structural racism of the archive, and, to a certain extent, the racist basis of the national history that is predicated solely on particular types of records (or their lack). This symbolic reversal is incredibly powerful, seeking to enable a connection pre-slavery, a way of allowing a community to connect with an origin point, in a way a revelatory ability to see 'beyond' or 'before' slavery, and in some ways to obviate the violence of the past centuries by offering an historical short-circuit that avoids it. Hence the shows have a public historical purpose insofar as they challenge the normative assumptions of 'history' and suggest in their place a different way of accessing and reclaiming the past. They show that marginalisation and effacement is a choice, and that history which repeats forgetting is simply underlines this. They argue that genetic knowledge is a special type of information that can challenge and change received understanding of the past, making then and now more complex. The new information serves to undermine the credibility of the old information, insofar as it allows a community to see how its 'history' has been 'constructed' for it using particular means (textual information and intentional forgetting). In place of this 'old' history is a new means of connecting with a past.

Gates, Jr's shows do emphasise genetic complexity. *Finding Your Roots* is particularly interested in the diversity of background which can be seen when investigating the genetic genealogy of modern figures. The historiographical point of this show is to suggest that our contemporary understanding of identity may not square with the genetic reality. The series participates in a kind of genomic literalism, suggesting that genetic ethnic background can change contemporary identity. At the same time it also creates dissonance in modern identity, challenging the normative and undermining the idea of totality in selfhood and demonstrating what Keith Wailoo calls 'the impossible tension between genetic claims and personal modes of identification'.[91] The DNA work also gives much longer-term heritage in this show, tracing back sometimes hundreds of years to locations around the world including modern-day Gabon, Cameroon, Korea, and Ireland. The work that is done ensures a firmer sense of connection to the past, as well as a clearer concept of identity in the present: participant Sanjay Gupta claims 'I feel more Indian today, and I feel greater pride in my heritage as well [...] I've always been proud to be Indian, but now I feel like I have a real story to tell'.[92] As this example shows, genetic information can prompt a shift in personal understanding of the past and of its relation to contemporary identity.

Notes

1 See Clare Hanson, *Genetics and the Literary Imagination* (Oxford: Oxford University Press, 2020).

2 A selection from the past 20 years: Nessa Carey, *The Epigenetics Revolution* (London: Icon Books, 2012), *Junk DNA* (London: Icon Books, 2015), and *Hacking the Code of Life* (London: Icon Books, 2019); Sidhartha Mukherjee, *The Gene: An Intimate History* (London: Vintage, 2017); Matt Ridley, *Genome: The Autobiography of a Species in 23 Chapters* (London: Fourth Estate, 1999); and *Nature Via Nurture: Genes, Experience, and What Makes Us Human* (London: Harper Perennial, 2004); George M. Church, *Regenesis* (London: Basic Books, 2012); Adam Rutherford, *A Brief History of Everyone Who Ever Lived* (London: Weiden and Nicolson, 2017); David Reich, *Who We Are and How We Got Here* (Oxford: Oxford University Press, 2018); Jennifer Doudna and Samuel Sternberg, *A Crack in Creation: The New Power to Control Evolution* (London: Vintage, 2018); Christine Kenneally, *The Invisible History of the Human Race: How DNA and History shape our identities and our futures* (London: Penguin, 2014).

3 See https://www.thefa.com/bootroom/resources/coaching/the-england-dna-what-do-you-need-to-know [accessed 24 August 2020].

4 Miguel Delaney, 'Lionel Messi Transfer', *Independent* 25 August 2020, https://www.independent.co.uk/sport/football/transfers/lionel-messi-barcelona-contract-latest-next-club-news-a9688591.html; PA, 'Racing Point launch appeal', *Guardian* 12 August 2020, https://www.theguardian.com/sport/2020/aug/12/racing-point-launch-appeal-against-mercedes-car-copying-fine-formula-one-motor-racing; 'Sir Bill Beaumont', NCA Rugby, https://www.ncarugby.com/national-2-north/beaumont-community-game-is-the-dna-of-rugby/ [all accessed 26 August 2020].

5 'For Conservatives to have hope, Trump has to lose', *New York Times* 24 August 2020, shorturl.at/akZ46 [accessed 24 August 2020].

6 Ben Quinn, 'Boris Johnson faces Tory rebellions on Brexit and Covid rules', *The Guardian* 11 September 2020, https://www.theguardian.com/politics/2020/sep/11/boris-johnson-tory-rebellions-brexit-covid-rules [accessed 21 September 2020].

7 'Sylvester: Step II', *Pitchfork*, 7 February 2021, https://pitchfork.com/reviews/albums/sylvester-step-ii/ [accessed 7 February 2021].

8 A community-created 'DNA Double Helix Discovery' has over 2000 supporters at https://ideas.lego.com/projects/5bd5311e-5078-4a7b-8763-e9cfd8a38c42 [accessed 24 August 2020].

9 Kurt F. Gothelf, 'LEGO-like DNA Structures', *Science* 338:6111 (2012), 1159-60.

10 Norbert Pardi, Michael J. Hogan, Frederick W. Porter and Drew Weissman, 'mRNA vaccines – a new era in vaccinology', *National Review of Drug Discovery* 17:4 (2018), 261-79.

11 'Understanding mRNA COVID-19 Vaccines', Centers for Disease Control and Prevention, 18 December 2020, https://www.cdc.gov/coronavirus/2019-ncov/vaccines/different-vaccines/mrna.html [accessed 19 January 2021].

12 Takehiro Ura, Akio Yamashita, Nobuhisa Mizuki, Kenji Okuda and Masaru Shimada, 'New vaccine production platforms used in developing SARS-CoV-2 vaccine candidates', *Vaccine* 39:2 (2021), 197-201 (200).

13 Jeremie Richard, 'Gene-mapping champion Iceland leads the way in COVID sequencing', *Medical Xpress*, 16 January 2021, https://medicalxpress.com/news/2021-01-gene-mapping-champion-iceland-covid-sequencing.html [accessed 19 January 2021].

14 Lisa A. Jackson et al, 'An mRNA Vaccine against SARS-CoV-2 – Preliminary report', *New England Journal of Medicine*, 2020; 383:1920-1931.

15 Regulatory approval in the UK and other information on the UK Government webpage published 2 December 2020, https://www.gov.uk/government/publications/regulatory-approval-of-pfizer-biontech-vaccine-for-covid-19 [accessed 19 January 2021].

16 UK Government Press release, 8 January 2021, https://www.gov.uk/government/news/moderna-vaccine-becomes-third-covid-19-vaccine-approved-by-uk-regulator [accessed 19 January 2021].

17 Megan Molteni, 'Why it's a big deal if the first Covid vaccine is 'Genetic'', *Wired*, 11 October 2020, https://www.wired.com/story/why-its-a-big-deal-if-the-first-covid-vaccine-is-genetic/ [accessed 19 January 2021].

18 Jason Murdock, 'Why mRNA COVID vaccines can't change your DNA', *Newsweek* 5 January 2021, https://www.newsweek.com/covid-coronavirus-mrna-vaccines-human-dna-conspiracy-theory-fact-check-1558962 [accessed 19 January 2021].

19 Flora Carmicheal and Jack Goodman, 'Vaccine rumours debunked', BBC news, 2 December 2020, https://www.bbc.co.uk/news/54893437 [accessed 19 January 2021].

20 Katie Weston, 'Doctors fear 'fake news'', *Mail Online* 15 January 2021, https://www.dailymail.co.uk/news/article-9150489/Doctors-fear-fake-news-causing-BAME-people-reject-Covid-vaccine.html [accessed 25 January 2021].

21 Over 70 Jewish doctors signed a letter responding to rumours about infertility and the Vatican intervened to say that vaccines were 'morally acceptable' despite the use in original design of aborted fetal cells, Sandy Rashty, 'More than 70 Doctors', *Jewish News* 25 January 2021, https://jewishnews.timesofisrael.com/jewish-doctors-say-absolutely-no-evidence-covid-19-vaccine-causes-infertility/ and Courtney Mares, 'Vatican says COVID-19 vaccines', *The Catholic World Report*, 21 December 2020, https://www.catholicworldreport.com/2020/12/21/vatican-says-covid-19-vaccines-morally-acceptable-when-no-alternatives-are-available/ [both accessed 25 January 2021].

22 Bruno J. Strasser, 'Who cares about the double helix?', *Nature* 422 (2003), 803-4.

23 Information from https://www.congress.gov/bill/108th-congress/senate-concurrent-resolution/10/all-info [accessed 18 May 2020].

24 'About DNA Day', https://www.genome.gov/dna-day/about [accessed 18 May 2020]

25 PA, 'Twitter troll' bombarded Labour MP Stella Creasy with abuse, court hears', *Guardian* 19 May 2014, https://www.theguardian.com/uk-news/2014/may/19/twitter-labour-mp-stella-creasy-court [accessed 19 May 2020].

26 See Svetlana Boym, *The Future of Nostalgia* (Basic Books, 2001) and Pierre Nora, *Realms of Memory*, trans. Arthur Goldhammer (New York, NY: Columbia Univesity Press, 1996).

27 Subhara Das, 'Francis Galton and the History of Eugenics at UCL', 22 October 2015, https://blogs.ucl.ac.uk/museums/2015/10/22/francis-galton-and-the-history-of-eugenics-at-ucl/ [accessed 12 June 2020].

28 Josh Gabbatiss, 'James Watson: The most controversial statements made by the father of DNA', *The Independent* 13 January 2019, https://www.independent.co.uk/news/science/james-watson-racism-sexism-dna-race-intelligence-genetics-double-helix-a8725556.html#gsc.tab=0 [accessed 5 June 2020].

29 'Decoding Watson', PBS, 1 February 2019, https://www.pbs.org/video/decoding-watson-ua6jjx/; N'dea Yancey-Bragg, 'Lab revokes honorary titles for Nobel Prize winner James Watson after repeated racist comments', *USA Today* 14 December 2019, https://eu.usatoday.com/story/news/nation/2019/01/13/dna-pioneer-james-watson-honors-racist-comments/2565503002/ [both accessed 5 June 2020].

30 Clarissa Place, "Abhorrent' road name to be changed to honour work of female scientist', *Norwich Evening News*, 26 June 2020, https://www.eveningnews24.co.uk/news/health/rosalind-franklin-honoured-in-norwich-road-name-change-1-6717348 [accessed 14 July 2020].

31 Josh Gattatiss, 'James Watson'; Rachel Feltman, 'The father of DNA is selling his Nobel Prize because everyone thinks he's racist', *Washington Post* 1 December 2014, https://www.washingtonpost.com/news/speaking-of-science/wp/2014/12/01/nows-your-chance-to-buy-james-watsons-nobel-prize-because-racism/; Philip Sherwell, 'DNA father James Watson's "holy grail" request', *The Telegraph* 20 May 2009, https://www.telegraph.co.uk/news/worldnews/northamerica/usa/5300883/DNA-father-James-Watsons-holy-grail-request.html [all accessed 5 June 2020].

32 David A. Wheeler et al, 'The complete genome of an individual by massively parallel DNA sequencing', *Nature* 452 (2008), 872-6. The first genome fully described was that of the HGP, largely based on Craig Venter, the CEO of Celera Genomics.

33 It is available at ftp://ftp.ncbi.nih.gov/pub/TraceDB/Personal_Genomics/Watson/ [accessed 5 June 2020].

34 Wheeler et al, 'The complete genome', p. 872.

35 Item description at https://www.christies.com/lotfinder/books-manuscripts/watson-james-dewey-nobel-prize-medal-in-5857953-details.aspx [accessed 5 June 2020].

36 Andrew Griffin, 'Arsenal owner Alisher Usmanov hands Nobel Prize back to disgraced DNA scientist James Watson straight after buying it off him', *The Independent*, 9 December 2014, https://www.independent.co.uk/news/science/arsenal-owner-alisher-usmanov-hands-nobel-prize-back-to-disgraced-dna-scientist-james-watson-9912725.html [accessed 5 June 2020].

37 See for instance *The Story of Post-Modernism* (Chichester: John Wiley, 2011).

38 'Double Helix Sculpture', https://www.saatchiart.com/art/Sculpture-Double-Helix/94935/3152950/view [accessed 29 May 2020].

39 https://m-tec.uk.com/projects/debenhams/ [accessed 29 May 2020].

40 https://olafureliasson.net/archive/artwork/WEK100857/umschreibung#slideshow [accessed 29 May 2020].

41 Tina Egstrom, 'DNA Inspired Art takes over London', GuideLondon 20 July 2015, https://www.guidelondon.org.uk/blog/around-london/dna-inspired-art-takes-over-london/ [accessed 9 June 2020].

42 Photos of most of the pieces are collected under the hashtag 'DNA Trail' on Twitter: https://twitter.com/hashtag/DNATrail [accessed 9 June 2020].

43 https://someoneinlondon.com/projects/dna-inspires-someones-7ft-london-sculptures [accessed 9 June 2020].

44 Science Museum Group, Crick and Watson's DNA molecular model, 1977-310 (Science Museum Group Collection Online), https://collection.sciencemuseumgroup.org.uk/objects/co146411/crick-and-watsons-dna-molecular-model-molecular-model [accessed 10 June 2020].

45 https://collection.sciencemuseumgroup.org.uk/search/gallery/making-the-modern-world-gallery/museum/science-museum [accessed 10 June 2020].

46 https://www.sciencemuseum.org.uk/see-and-do/making-modern-world [accessed 10 June 2020].

47 Francis Crick, *What Mad Pursuit* (New York, NY: Perseus Books, 1988), p. 65.

48 Rita Felski, 'Context Stinks!', *New Literary History* 42:4 (2011), 573-91 (p. 579).

49 The source for some of this information are not clear, see 'Antony Barrington Brown', *The Telegraph* 14 February 2012, https://www.telegraph.co.uk/news/obituaries/culture-obituaries/9082749/Antony-Barrington-Brown.html [accessed 10 June 2020].

50 The image was bought by the National Portrait Gallery, London, in 1994, https://www.npg.org.uk/collections/search/portrait/mw13740/James-Dewey-Watson-Francis-Harry-Compton-Crick [accessed 10 June 2020].

51 1953 article cite.

52 R. E. Franklin and R. G. Gosling, 'Molecular configuration in sodium thymonucleate' *Nature* 171 (1953), 740–741.

53 Brenda Maddox, 'The double helix and the "wronged heroine"', *Nature* 421 (2003), 407-8 (p. 408).

54 *The Double Helix* (New York: Touchstone Press, 1996), p. 181, originally published in 1968, p. 181.

55 *Rosalind Franklin and DNA* (New York: W.W. Norton, 2000), reprinted from 1975.

56 Jasper Hammill, 'Plaque off', *The Sun*, 10 October 2017, https://www.thesun.co.uk/tech/4650825/feminists-vandalise-blue-plaque-to-dna-discovery-scientists-francis-crick-and-james-watson-in-tribute-to-forgotten-female-researcher-rosalind-franklin/ [accessed 19 May 2020].

57 Press release, 'Name of British built rover revealed', 7 February 2019, https://www.gov.uk/government/news/name-of-british-built-mars-rover-revealed [accessed 27 July 2020].

58 Josh Gabbatis, 'New £50 note', *The Independent* 2 November 2018, https://www.independent.co.uk/news/science/new-50-note-scientists-most-likely-stephen-hawking-ada-lovelace-rosalind-franklin-alan-turing-a8615001.html [accessed 19 May 2020].

59 Michael Billington, '*Photograph 51* review', *The Guardian*, 14 September 2015, https://www.theguardian.com/stage/2015/sep/14/nicole-kidman-photograph-51-noel-coward-theatre-rosalind-franklin-review; Steve Connor, 'Nicole Kidman in *Photograph 51*', *The Independent*, 21 September 2015, https://www.independent.co.uk/arts-entertainment/theatre-dance/features/nicole-kidman-in-photograph-51-a-new-play-explores-rivalries-over-dna-10511581.html; Ben Brantley, 'Review: In *Photograph 51* Nicole Kidman is a Steely DNA Scientist', *New York Times*, 14 September 2015, https://www.nytimes.com/2015/09/15/theater/review-in-photograph-51-nicole-kidman-is-a-steely-dna-scientist.html [all accessed 19 May 2020].

60 *Photograph 51* (London: Oberon books, 2015), p. 7.

61 Richard Buckley, Mathew Morris, Jo Appleby, Turi King, Deirdre O'Sullivan and Lin Foxhall, '"The king in the car park": new light on the death and burial of Richard III in the Grey Friars church, Leicester, in 1485', *Antiquity*, 87 (2013), 519-538.

62 Turi E. King et al, 'Identification of the Remains of King Richard III', *Nature Communications* 5: 5631, 2 December 2014, http://www.nature.com/ncomms/2014/141202/ncomms6631/full/ncomms6631.html

63 Philippa Langley and Michael Jones, *The King's Grave* (London: John Murray, 2013), Kindle edition paragraph 4.

64 John F. Burns, 'Bones under parking lot belonged to Richard III', *New York Times*, 4 February 2013, https://www.nytimes.com/2013/02/05/world/europe/richard-the-third-bones.html [accessed 22 May 2020].

65 Matthew Morris and Richard Buckley, *The King Under the Car Park* (Leicester: University of Leicester, 2013).

66 'Richard III dig: DNA confirms bones are king's', *BBC* 4 February 2013, https://www.bbc.co.uk/news/uk-england-leicestershire-21063882 [accessed 22 May 2020]; Deborah Netburn, 'King Richard III's DNA opens a door to a new historical mystery', *Los Angeles Times* 2 December 2014, https://www.latimes.com/science/sciencenow/la-sci-sn-king-richard-iii-dna-20141201-story.html [accessed 22 May 2020].

67 Nicholas Wade, 'Tracing a Royal Y Chromosone', *New York Times* 11 February 2013, https://www.nytimes.com/2013/02/12/science/more-dna-tests-to-confirm-skeleton-is-richard-iiis.html [accessed 12 June 2020].

68 *The King in the Car Park*, dir: Darlow Smithson, Channel 4, 4 February 2018.

69 Andrew Marszal, '*The King in the Car Park*, Review', *The Telegraph*, 4 February 2018, https://www.telegraph.co.uk/culture/tvandradio/9848456/Richard-III-the-King-in-the-Car-Park-Channel-4-review.html [accessed 26 February 2019].

70 Richard Toon and Laurie Stone, 'Game of Thrones: Richard III and the Creation of Cultural Heritage' in *Studies in Forensic Biohistory*, ed. Christopher M. Stojanowski and William N. Duncan (Cambridge: Cambridge University Press, 2016), pp. 43-67 (p. 59).

71 Lucia Marchini, 'Richard revisited', *Current Archaeology* 12 September 2014, https://www.archaeology.co.uk/articles/reviews/richard-revisited.htm [accessed 22 May 2020].

72 Carly Hilts, 'Return of the King', *Current Archaeology* 2 March 2015, https://www.archaeology.co.uk/articles/features/return-of-the-king-richard-iiis-remains-are-taken-to-leicester-cathedral.htm [accessed 22 May 2020].

73 'Richard III: More than 5000 people visit Leicester Cathedral coffin', *BBC* 23 March 2015, https://www.bbc.co.uk/news/uk-england-leicestershire-32014296 [accessed 22 May 2020].

74 There is a detailed account of the service and the day's celebrations here: Victoria Ward, 'Reburial of Richard III – as it happened', *The Telegraph* 26 March 2015, https://www.telegraph.co.uk/news/earth/environment/archaeology/11495617/Reburial-of-Richard-III-As-it-happened.html [accessed 22 May 2020].

75 Cited in Ward, 'Reburial of Richard III'.

76 Maev Kennedy, 'Benedict Cumberbatch is related to Richard III, scientists say', *The Guardian*, 25 March 2015, https://www.theguardian.com/uk-news/2015/mar/25/benedict-cumberbatch-is-related-to-richard-iii-scientists-say [accessed 21 May 2020].

77 King Richard III Visitor Centre, https://kriii.com/about-the-centre/dynasty-death-and-discovery/ [accessed 21 May 2020].

78 *Thomas Jefferson and Sally Hemings: An American Controversy* (Charlottesville and London: University of Virginia Press, 1997).

79 Eugene A. Foster, et al, 'Jefferson fathered slave's last child', *Nature* 396 (1998), 27-8.

80 Christine Kenneally, *The Invisible History of the Human Race* (New York: Viking Penguin, 2014), p. 227.

81 Howard Schuman and Amy Corning, 'The roots of collective memory: Public knowledge of Sally Hemings and Thomas Jefferson', *Memory Studies* 4:2 (2010), 134-53.

82 *Free Some Day* (Chapel Hill, NC: University of North Carolina Press, 2002); *The Hemingses of Monticello: An American Family* (New York: Norton, 2008).

83 Catherine Kerrison argues for a focus on Hemings herself, rather than as relational to Jefferson, in 'Sally Hemings' in *A Companion to Thomas Jefferson*, ed. Francis D. Cogliano (Oxford: Blackwell Publishing, 2012), pp. 284-300.

84 https://www.monticello.org/slavery-at-monticello/about/visiting-our-exhibition [accessed 11 June 2020].

85 *Jefferson's Children* (New York: Random House, 2002).

86 *Thomas Jefferson and Sally Hemings: An American Controversy* (Charlottesville and London: University of Virginia Press, 1998), Kindle edition, 'Author's Note', paragraph 1.

87 Michael L. Blakey, 'Bioarchaeology of the African Diaspora in the Americas: Its Origins and Scope', *Annual Review of Anthropology* 30 (2001), 387-422. See also Rick Kittles and Charmaine Royal, 'The Genetics of African Americans', in *Genetic Nature/ Culture* ed. Alan H. Goodman, Deborah Heath, and M. Susan Lindee (Berkley, CA: University of California Press, 2003), pp. 219-32.

88 Carey Goldberg, 'DNA offers link to Black History', *New York Times*, 28 August 2000, https://www.nytimes.com/2000/08/28/us/dna-offers-link-to-black-history.html [accessed 3 July 2020].

89 https://africanancestry.com/home/ accessed 3 July 2020.

90 Kindle edition, 'Introduction' paragraph 10.

91 Keith Wailoo, 'Genes and the Problem of Historical Identity', Wailoo, Nelson and Lee, p. 18.

92 *Finding Your Roots* (Chapel Hill, NC: University of North Carolina Press, 2014), p. 227.

2
PRACTICE

This chapter looks in close detail at the ways in which genetic data and analysis is changing the practice of history. In Chapter 1, we considered some of the ways that genetic information has affected and effected commemorative practices in the United Kingdom and around the world. We also looked at how information derived from DNA might change historical understanding and awareness. In this chapter, we begin to see some of the ways that new genetic information might challenge historical modes of enquiry and shift thinking about interpretation of the past. The chapter considers DNA as historical 'practice', that is, as a mode of investigation of the past and a means for understanding and narrating events. In particular, this chapter focuses on practice related to the investigation and interpretation of Ancient DNA (aDNA) found in samples taken from archaeological sites.

Ancient DNA analysis involves the consideration of ancient human, animal, pathogen and plant samples, as well as analysis of microbiomes found in dental pulp, earth, and fossilised faeces.[1] The rhetoric around what David Reich calls the 'ancient DNA revolution', and Kristian Kristiansen has termed 'the third scientific revolution', suggests an entirely new way of thinking about the past and our approach to it.[2] Whilst aDNA had been used in interdisciplinary historical investigation previously, the explosion in data produced from 2010 onwards has meant that 'historians and archaeologists are grappling with a new scientific technique'.[3] Descriptions of the area argue that the expansion of information is changing *what* we know about the past: 'Findings from aDNA research are currently transforming our understanding of human history at an ever-increasing pace [...] scientists are racing to apply the work to answer questions about human evolution and history that would have been unfathomable just a few years ago'.[4] Furthermore research groups themselves make wide-ranging claims for the impact of their work on *how* we know the past: 'Genome-wide analysis of ancient DNA

DOI: 10.4324/9781003052975-3

has emerged as a transformative technology for studying prehistory, providing information that is comparable in power to archaeology and linguistics'.[5]

What does this mean for the practice of history, and for historians more generally? The rapid expansion of aDNA research and concomitant increase of data raises significant disciplinary and epistemological questions that are discussed throughout this book.[6] The surge of work on aDNA since 2010, and the 'historical' information it has generated, might be considered in depth as a means for reflecting upon the shifts that genetic science are effecting on historical knowledge and practice.[7] On the one hand, historians must work out a mode of engagement with this work that allows critique and interrogation of the basic concepts and language that is being used; on the other, they need to recognise the field as useful and develop tools for collaborative working.[8] Therefore, this chapter outlines the changes to historical practice suggested by work on aDNA. Firstly, it outlines and critically discusses some of the key aspects of work in the area, reflecting upon how we might conceive of the analysis of aDNA material as an historical praxis. This aspect of the chapter considers how aDNA work might impact upon the production of, and indeed definition of, historical data and knowledge. Secondly, the chapter considers the implications aDNA research has for disciplinarity. Genomics is beginning to situate itself as an historical practice, suggesting that researchers in the area might be considered historians, discovering extra sources, presenting innovative evidence, and outlining new readings of the past. This could be seen as the beginning of an historiography of aDNA, a reflection upon the challenge to historical 'knowledge' represented by the practices of the geneticist-historian. Throughout it is important to recognise the duality of aDNA as something that is imaginatively conceived of both as material thing and way of knowing, a tool *and* a mode of knowledge.

Genomic Histories: Ancient DNA Analysis and Historical Practice

(Ancient) Geneticists and History

Investigation of human development through the genetic evaluation of archaeological samples began in the mid-1980s.[9] The first work in this area sought to confirm that it was possible to extract DNA from ancient samples.[10] Thereafter remains were subject to an increasing diversity of investigation, and the modes of extraction and analysis became more sophisticated.[11] The approach was increasingly hailed as important and useful, having 'significant implications for the study of past populations as it provides a source of primary evidence to add to the indirect evidence gained from modern population genetics, and from linguistic, cultural and anthropological sources'.[12] This latter comment illustrates how, from the beginning, aDNA work was considered something that would produce *new* knowledge in the form of 'primary evidence'.

The development of techniques quickly refined the data available.[13] In 1992 Terrence Brown and Keri Brown introduced 'what promises to be a major area of archaeological science' and suggested that Ancient DNA would be useful for archaeologists studying migration, kinship, and disease.[14] In particular anthropologists were 'quick to adopt these new techniques for the production of previously *unobtainable* data' (my emphasis) seeing 'the potential to add greatly to our understanding of human/ primate evolution and history'; this led to important work on ethics, population movement, and evolution.[15] Ancient DNA analysis was used on ancient manuscripts and on samples found around the world at archaeological sites.[16] It 'was seen as providing a unique viewpoint to the temporal dynamics of human evolution'.[17] That said, there were still significant variations in practice which called into question the robustness of results.[18]

Between 2000 and 2010 the field changed significantly in terms of approach, consistency of results, and profile. The development of 'next generation' sequencing in 2005 made generating information quicker, more consistent, and cheaper.[19] Contemporary investigation means using sampling techniques that consider longer DNA fragments and produce significantly larger datasets.[20] The information generated is useful when aligned with data about migration and population change. In 2010 genome sequences were published of ancient humans, including Neanderthals, the newly discovered Denisovans, and a 4000-year old 'Paleo-Eskimo'.[21] There has subsequently been a huge expansion in the number of studies, with far-reaching work published regularly in major journals such as *Nature* and *Science*.[22] In a 2018 (non-exhaustive) overview Pontus Skoglund and Iain Mathieson listed 100 publications produced during the 'the first decade' of new work (and their account excludes the huge amount of work on non-human (plant, animal, pathogen) aDNA).[23] A 2018 article in *Nature* suggests that 1300 'ancient' genomes have been sequenced since 2010.[24]

After 2010, with the presentation of the Neanderthal genome and the revelation of a previously unknown group related to them, known as Denisovans, work in the area approached 'exponential growth' and in many areas to begin to substantially encroach upon research in archaeology and history.[25] Many commentators described 2010 as a 'the first year of ancient human genomics'. The rhetoric around ancient DNA work (and reporting upon it) is caught up in a sense of profound change from then onward, associated with a sense of the expansion of new data.[26] In a 2017 review article surveying the field of paleogenomics Michela Leonardi and her co-authors reported that 'more than 500 ancient humans have been analysed for genome-wide sequence variation' with 'nearly 80 ancient human genomes [...] characterized'.[27] This expansion is accelerating, with Marciniak and Perry claiming an 'ancient genome explosion'.[28] Leonardi et al.'s account of work on aDNA lists analysis of humans from 200 to 45000 years BP and studies of Neanderthal and Denisovan samples from 40000 to 50000 years BP. They cite as part of the wider field the genetic investigation of pathogens from 150 years to 4887 BP and mammals such as horses and wolves from 100 years BP to 780000 years BP.[29]

Ancient DNA analysis gives perspective on evolution of animals, pathogens, and plants, even those that are now extinct.[30] It has informed discussion of contemporary ecosystems, as Beth Shapiro reminds us: 'Ancient DNA has turned out to be a powerful technique for leaning about the evolutionary processes that shaped existing biodiversity'.[31] Ancient DNA depends, at present, on samples found in usable condition, which means much of the work has been undertaken on materials from 'cold and temperate world regions', itself problematically focussing study on particular areas, although this is changing as the technology becomes more sophisticated.[32] Contemporary studies of aDNA material make up a diverse area, robust and evolving.[33] Articles consider how change in material culture 'imprints' genomic change over millennia or focus on particular temporal moments.[34] Contributions seek to prove points, add information, or report new data. They present new techniques or discuss the ethics of work on historical samples.[35] They engage in debate about matrilineal dynasties or expand knowledge of little-known communities.[36] Research groups work ingeniously to reconstruct unknown genomic histories.[37] They situate themselves as augmenting and broadening knowledge.[38] Links to contemporary populations are worked through and conclusions drawn.[39] There is an increasing self-consciousness about terminology and bias in investigation.[40]

The expansion in information and analysis means that aDNA investigation can seemingly transform knowledge of recent historical events.[41] It is changing how historians might study the eighteenth-century slave trade, or the burial practices of seventh-century Germanic tribes, or the movement of peoples across Europe in the sixth century.[42] In these instances, genetic data used in tandem with established techniques enables a wider perspective and new knowledge: 'we directly tested historical and archeological hypotheses'; 'we used genome-wide data to trace the origins of three enslaved Africans'.[43] As with this latter example, often this work is able to redress problematic lacuna, as Jada Benn Torres argues: 'this type of approach can help to fill gaps in knowledge that were created from the destructive effects of colonization'.[44] The institutions of history – the state, the archive, the library – can be challenged by this work, and the production of historical knowledge shifted.

Breakthroughs, announcements and innovations in this area account for some of the highest profile archaeological and genetics stories over the past 10 years. Since the wide media coverage given to the Richard III case in 2013–2014, aDNA investigation has been widely reported.[45] Important moments include the controversy around 'Cheddar Man' (2018, see Chapter 3), reporting on the 'Beaker people' (2018), the Anzick-1 and Kennewick man cases (2014–15), the revisionism about 'Egtved girl' (2015), and debate around 'Mungo Man' (2016).[46] Each event provoked public debate about the nature of memory, the ethics of commemoration, and the role of genetic science in presenting knowledge of the past.[47] In the cases of the remains of Anzick-1, ceremonially reburied by Montana tribal elders in 2015, or Richard III, given a public state reinternment in the United Kingdom in 2015, investigation of DNA samples becomes directly related to particular types of commemorative practice. In the case of 'Egtved girl', 'Cheddar Man', or the 'Beaker' people, genetic investigation challenged long-held assumptions about nationhood and race (see Chapter 3). For instance, 'Egtved Girl', a longtime symbol for Danish identity and historical stability was proven to be 'from a place outside present day Denmark' and the case was used to highlight Bronze age mobility and migration.[48] Analysis of 'Cheddar Man', some of the oldest known hominid remains in Europe, suggested that his skin pigmentation was dark. Ancient DNA investigation is neither politically neutral nor culturally inert. It intervenes into historical and cultural arguments about identity, nationhood, race, biological difference, and migration. It can interrogate longstanding memorial tropes, prompt revisionary accounts, and allow the rethinking of events.

Similarly public interest in the new wave of ancient geneticists is high, with leading figures such as Eske Willerslev and David Reich being profiled by influential media outlets or writing popular science books about aDNA.[49] Accounts of ancient DNA work in the science press emphasise the transformative and revelatory aspect of the work. The BBC reports how 'ancient DNA is transforming our view of the past'.[50] Other media outlets echo this language of change: *Atlantic* reports 'Ancient DNA is Rewriting Human (and Neanderthal) History'; the *Independent* that 'DNA analysis of humans from the Bronze Age has *revealed* unprecedented insights'; the *Washington Post* 'Neanderthal microbes reveal surprises'; and the *Daily Mail* 'DNA from 8000-year-old skeletons reveals' information about contemporary populations.[51] Work on aDNA is extremely high profile, and is ensuring a public sense that the past can be changed and reconfigured through the interventions of genetic work.

The mode of reporting of much of the aDNA work is revelatory, as historical 'truths' are challenged or changed by this new information, still regularly reported as raising 'hopes that it could transform our understanding of how our predecessors behaved'.[52] In a 2015 article entitled 'Ancient DNA cracks puzzle of Basque origins', for instance, the BBC reported that 'DNA from ancient remains seems to have solved the puzzle of one of Europe's most enigmatic people: the Basques'.[53] The editorial team outlined how work on genomes of Stone Age skeletons from northern Spain suggested that these remains from 3500 to 5500 years ago are related to the contemporary population.[54] This type of reporting articulates the importance of DNA work in revealing new connections and opening up historical truths. Often the rhetoric is about that which has been lost, ignored, or is unseen; Alice Roberts writes that 'Ancient DNA bears clues to forgotten journeys'.[55] Writing about aDNA investigation, then, often seems to suggest that it is a new mode of knowing the past, introducing important evidence that had been hitherto missed and giving clarity to historical events. This popular repetition of the revelatory rhetoric of scientific papers (discussed below) demonstrates the impact of this language in buttressing an epistemological framework. In particular it demonstrates how being 'postgenomic' might be something constructed culturally. In this media telling understanding prehistory has become dependent on postgenomic awareness; the sophistication of our contemporary understanding of the past is engendered by our new knowledge.

Ancient DNA Research as Historical Practice

Given this shift in the public historical imaginary to incorporate aDNA as both a new set of evidence and a new way of knowing the past, it is important to recognise how the studies themselves are articulating their practice. Contributions to the aDNA debate present accounts of the past using genetic and genomic data as evidence to make assertions and to knit narrative together. Work in the field often uses 'genetic history' comfortably as a term that gives form and completeness to the data. There is an assumption of the chronological arrangement of genetic interrelationship, 'genetic history' as a field of knowledge about the status of DNA in time. More importantly, and increasingly self-consciously, aDNA research is contributing historical data and historiographic innovation.[56] The research teams are adding to the way that the past is interrogated and what kind of information is produced. This is a clear methodological as well as epistemic engagement with historical practice. 'History' is undertaken in these articles, as an entire model of the past and the way it might be narrated and investigated is being propounded.[57] Each paper provides a complex interaction of research work and analysis, and each might be investigated thoroughly as a case-study to outline the biases, institutional contexts, and purposes of contemporary aDNA investigation.[58] Two recent examples of such 'genetic history', considered below, demonstrate the ways in which aDNA research presents historical practice. This allows us to contemplate the model of historical investigation that is being propounded and to make suggestions about the modes of knowing that are revealed, particularly in language usage and information articulation.

Qiaomei Fu et al.'s 2016 'The genetic history of Ice Age Europe' is an overview of 51 Eurasians from 45000 to 7000 years.[59] DNA was extracted from human remains and processed into sequencing libraries according to standard protocols. One problem with aDNA is that 'the vast majority of the DNA extracted from most specimens is of microbial origin' (p. 200), and this makes sequencing inefficient In order to deal with this, Fu et al. augmented the DNA libraries through innovative techniques. This allows them to then use 'shotgun sequencing' in order to identify what part of the DNA maps onto the human genome. Having taken care to avoid contamination with modern DNA, the analysis datasets are constructed.[60] Much of the work entails comparing the created dataset with known sequencing data (in this case, from previous work published on archaic and modern humans). Once the final dataset is produced the team use various methods to compare it with known information about population and genetic makeup. The information is compared with published statistic to make assertions about natural selection, the impact of climate, and genetic changes:

Modern humans arrived in Europe ~45,000 years ago, but little is known about their genetic composition before the start of farming ~8,500 years ago. Here we analyse genome-wide data from 51 Eurasians from ~45,000–7,000 years ago (p. 200).

The piece contributes an account of some 38000 years via the interpretation of aDNA sample and the genetic make-up of ancient individuals. It adds to knowledge by providing a new insight (previously genome-wide information from only 4 'Upper Palaeothic individuals from Europe' existed). The data generated creates a source: 'This data set provides an unprecedented opportunity to study the population history of Upper Palaeolithic Europe over more than 30000 years' (p. 202). The human over 30000 years can be understood, narrated, and interpreted via the genetic information gathered from 51 individuals. This work creates a 'genetic history' which is a separate account from those given by other disciplines:

> In order not to prejudice any association between genetic and archaeological groupings among the individuals studied, we first allowed the genetic data alone to drive the groupings of the specimens, and only afterward examined their associations with archaeological cultural complexes' (p. 202).

The research group come to particular conclusions after considering the data that is generated: 'These results document how population turnover and migration have been recurring themes of European prehistory [...] We show that the population history of pre-Neolithic Europe was complex in several respects' (pp. 200, 204). The conclusions drawn relate mainly to population development: 'the appearance of the Villabruna Cluster may reflect migrations or population shifts within Europe at the end of the Ice Age [...] One scenario that could explain these patterns is a population expansion from southeastern European or west Asian refugia after the Glacial Maximum' (p. 204). The authors make interventions into debate about the diminution of Neanderthal ancestry in the European genome (p. 201). Furthermore, what is not-known will impact upon what is: 'An important direction for future work will be to generate similar ancient DNA data from southeastern Europe and the Near East to arrive at a more complete picture of the Upper Palaeolithic population history of western Eurasia' (p. 204). This article presents its information and its findings extremely clearly as an intervention into a continuum of knowledge that is, importantly, unencumbered by disciplinary boundaries. The article is useful for archaeologists, geneticists, anthropologists, paleogeneticists, population geneticists and many others. The type of knowledge it is producing is hard information, in the form of the datasets, and interpretation, in the form of the analysis of the results.

Produced by a group of 64 scholars around the world, the 'history' presented here is intensively collaborative. This work could not be done individually and is the collation of multiple contributions. Such research reflects a commitment to transnational networking, the expansion of knowledge, and to a particular historical revisionism. The initial collection of data from the past is undertaken and then modelled according to the most up to date, modern techniques. A 'genetic history' here is the linking of data through geography (Europe) and temporality (~45000–7000 years ago), and the analysis of such information leads to a set of conclusions. The data is interrogated using robust techniques that have been developed over several years, and by deploying particular tools and software to 'read' the information.

On a bigger scale even than Fu et al., a research group numbering seventy-five authors published ambitious work in *Nature* in May 2018 that sequenced the genomes of 137 ancient humans.[61] The research scope was 'to understand the population genetic processes associated with the linguistic and cultural changes of the steppes after the Bronze Age migrations' (p. 369). The work considers the genomes of 137 ancient humans with coverage of *c.* 4000 years from 2500BCE to 1500 AD. The research intervenes into a number of debates including the development of particular languages and cultures, the scope and impact of the Justinian plague, and shifts in genetic ancestry, 'a process that continued well into historical times' (p. 374). The article has contributions to make for scholars of the ancient past and those of the early modern period. Similarly to Fu et al. the approach involves genomic sequencing, statistical comparison, biomathematical modelling, comparison of ancestry, and pathogen analysis. The article's material conclusions are found in the appendix pages of graphs, charts, and tables comparing data and analysing populations, including figures that 'model the entire ancient and present-day diversity of Inner Asia using the key ancestral groups' (p. 373); as well as identification of genetic and migratory information drawn from the data and demonstrated in the results. The article works through the rehearsal of technique and narrativising of analysis, prefacing the presentation of conclusions and discussion. The discussion is careful and collaborative: 'These findings are consistent with archaeological models' (p. 370); 'Our findings fit well with current insights from the historical linguistics of this region' (p. 373). Ancient DNA work can 'fit well' with current work; the data derived 'are consistent' with other investigation. This is a tool that affirms other modes of inquiry. The notion of integration and mutual interdependence of the three fields (linguistics, archaeology, genetics) is a common trope for such work.

Genomic history is demarcated here as a way of approaching the past. It is expansive – creating new datasets that span millennia – and transnational. The research networks involved in these articles are constructing a kind of history. The initial collection of data from the past materials is undertaken and then modelled according to the most up to date, modern techniques and using (often briefly outlined) contemporary historical and archaeological scholarship. A 'genetic history' is produced which is the linking of genetic data through geography and temporality (here ~45000–7000 years ago or 2500 BCE to 1500 AD), and the linked analysis of such information leading to a set of conclusions and suggestions. DNA is the source material that is being studied, and the implications it raises are then cross-referenced with other investigative models. The analysis is subject to ethical approval and peer review. It is 'big' and 'deep' and has a firm grasp of a particular developmental chronology (which it seeks to complicate). The focus is on development and difference between populations, on migration, on seeking patterns and trying to understand them. The historian-geneticist is the person or group generating, archiving, and reading the data. They are the ones creating the evidence and defining what 'evidence' is and how it will be read. This is an investigation of process and change, with a narrative attempt at an overview to tie it together. The intervention enables a deeper understanding of human development. This historical investigation is enabled by technology and computing power, supported by developments in communication and collaborative approaches. The historian-geneticist is the person or group generating, archiving, and reading the data. They are defining what 'evidence' is and how it will be read.

Ancient DNA Research: Revelation and Periodisation

Given the claims made, and the sophisticated data analysis being deployed, these articles often base their hypothesis-testing on relatively brief historical arguments. They present relatively few archaeological or historical sources, and the archaeological work is added into a narrative that is often one-dimensional. The hypotheses that are being tested are broadly sketched, taking in wide temporal and geographical locations. Research writing on aDNA tends to stress the innovation of technique and the 'new' quality of knowledge. Ancient DNA evidence is presented as transformative: 'the population movements that accompanied these events have previously been unknown, owing to the lack of ancient DNA evidence'; 'we report the first extensive investigation of Aboriginal Australian genomic diversity'.[62] The articles suggest that they are building on, responding to, and revising previous work; simultaneously they are adding to human knowledge and expanding our understanding. The 'transformative technology' here provides the crucial information to provide the resolution of historical indeterminacy. Methodologically key is the refutation and complication of various historical or archaeological hypotheses and inference. Ancient DNA analysis is presented as revelatory insofar as it sheds light into unconsidered places, opening up discussion that had been hampered by lack of information. Articles demonstrate the 'potential of ancient genomes to shed new light on European human prehistory' (p. 2), renewing samples that archaeology or history have ignored, failed with, or moved on from. So Iosif Lazaridis et al. present 22 'new' individuals through returning to old aDNA samples and reassembling individual genomes from the fragments.[63] The data-rich profiles that are produced hence allow for 'new historical interpretation'.[64] Ancient DNA investigation considers the human as historical data. It renders the physical – the shard of bone, for instance – into figures and then considers this information in a range of ways. Such work investigates humanness and migration over long swathes of time. This allows 'new historical interpretation' and new knowledge: 'Over the last decade, paleogenomics has been instrumental in settling long-disputed archaeological questions'.[65]

Le Monde wrote presciently in 2006 that ancient DNA sequencing introduced 'Origines de l'homme: une histoire à réinventer'.[66] The language of reinvention and revelation is key to the tone and thrust of aDNA investigation. The effect of the research is historically revisionist, as the *New York Times* notes they use: 'methods potent enough to inspire a wholesale revision of our knowledge about ancient peoples'.[67] As we have seen, scientific practice has regularly been mapped onto historical methodology to create an historiographic impetus. A rhetoric of revelation as a means of achieving more complex truth is foundational to the way this research presents its methodological nexus, even in the most moderate of texts: 'Employment of ancient human DNA analysis can partially *resolve* this serious problem concerning our understanding of recent human evolution in this geographic area' (my emphasis).[68] In a study of the 'poorly understood' African populations that contain the 'most ancient genetic lineages in humans', Joseph Pickrell et al. suggest: 'the anthropological and archaeological evidence for this hypothesis is contested [...] Genomic studies have the potential to shed new light on the history of the Khoisan'.[69] The 'new light' might untangle the knot of previous controversy. In fact aDNA work regularly uses a language of uncovering and bringing to light:

> Our analysis provides genetic evidence that hunter-gatherers settled Scandinavia via two routes. We reveal that the first Scandinavian farmers derive their ancestry from Anatolia 1000 years earlier than previously demonstrated.[70]

Studies assert the 'power of ancient DNA is that it offers a window into past biota and evolutionary processes that is inaccessible using DNA from living organisms or paleontological studies alone', presenting ancient DNA as revelatory (enabling a new way of seeing) and illuminating that which was formerly 'inaccessible'.[71] 'Reveal' is one of the most commonly used words in articles demonstrating the outcomes of particular investigative practice (along with 'insight', 'document', 'show', 'report'). This revelation enables clarity and direction, as the newly generated information brings us 'towards a clearer view'.[72] Such revelatory terms are used repeatedly in the titles of papers, lending them force, and are then repeated in wider science reporting.[73] In these aDNA papers the contemporary research group 'reveals' that which was waiting to be understood. They introduce a new, untold number of known unknowns to historical investigation. The language encourages the sense that aDNA enables a new way of understanding the past. The porting of this language of revelation into what is an investigation and writing of a type of 'history' illustrates further that there is an unarticulated theoretical foundation to much aDNA research. Language and practice combine to present a descriptor for knowledge.

An important defining idea here is 'ancient'. 'Ancient DNA' most commonly refers to the damaged state of the material sample. This is a technical idea that is becoming a defining term (popular press articles regularly simply refer to 'Ancient

DNA' as a practice).[74] 'Ancient DNA' is a keyword and a mode of definition. The articles move between geological, archaeological, and other types of more familiar temporal terminology for their dating and definitions.[75] If we look at three key articles that partly initiated the current expansion, all published in 2010, we can see that the idea of 'ancient' DNA research can apply to a range of findings.[76] The sample that Morten Rasmussen et al. analyse was taken from the permafrost and is '4750–2500 C^{14} yrs BP' (p. 757), dated according to radiocarbon conventions. The sample is elsewhere suggested to be c.4000 years old. The Denivosan sample used by David Reich is from an 'archaic' hominin contrasted with anatomically modern humans. It is dated to '50,000 to 30,000' years ago (p. 2053). The Neanderthal data worked on by Richard Green is dated via radiocarbon to '38,310 +/− 2,130' (p. 711) and is given other more conventional dates ('extinct late Pleistocene hominin', p. 722).[77] Each of these temporal and archaeological descriptors imposes meaning upon the data, and the use of a variety of dating conventions and terms complicates understanding. Whilst this language is conventional and multiple types of dating information are generally used, the impact of such terms are in need of interrogation.

Articles in the field might talk about 'ancient individuals' (a study considering DNA from 44 individuals 12000–1400 BCE); look at the shifts in DNA before and after European contact on Rapanui; consider 'one of the oldest fossils of anatomically modern humans from Europe'; or present an 'ancient DNA transect-through-time in Britain'[78]. The geographic, geological, temporal, genetic, paleogenomic language is precise. The ways in which the materials they study are located temporally differ depending on which particular scientific discourse or practice is being invoked. Yet throughout 'Ancient' becomes an overarching definition meaning to be temporally pre- something. The contaminating effect of modern DNA means that studies work hard to ensure purity of sampling and to differentiate between 'endogenous' and 'contaminating sequences' of modern DNA. This creates a material distinction between contemporary and ancient, physically distinguishing then and now, which is communicated widely in reporting on these findings. Periodising and separating terms such as 'modern', 'ancient', 'archaic' and 'prehistory' are ported into scientific papers, used technically but also with historical import.

This malleability is significant. 'Ancient DNA' is an historical descriptor, an account of data, an umbrella term for a subdiscipline, and a technical approach, simultaneously. The flexibility of the term suggests the construction of an episteme, in particular focussed upon reconfiguring knowledge of the human in time.[79] The use of 'ancient' to clarify the differences between time periods shows the investigations themselves investing in a particular temporality. Ancient DNA work undertakes to contrast 'ancient' with modern in order to make asrtions. Due to better and faster processing which allows for more information to be analysed, there is an opening up of access to a new 'resource'. The 'ancient' body becomes part of a mode of knowledge production – the conceptualising and testing of hypotheses, the demonstration of 'truth' through the consideration of assertion.

The 'ancient human genomes' become a resource, an archive of information to test concepts with, a library to read and interrogate. The archive created consists of the hundreds of genomes sequenced. This concept of the 'ancient' human archive is a conceptual device used by such studies to enable a particular type of investigation, drawing on Larsen's idea of the body as a kind of text and historical archive.[80] The human is defined by these studies in terms of its relation to then or now, temporally assigned meaning.

Ancient DNA as History

Many of these 'tools' for approaching and describing the materials of the past did not exist a decade ago; certainly the amount of data available has expanded hugely. What is clear from the literature on aDNA is that many research groups consider DNA to have been 'transformed into an historical source' to use Adam Rutherford's words.[81] Considering aDNA's move into being 'an historical source' demands a critical rethinking of approach, disciplinary boundaries, methodology and theory. It is crucial to understand what type of 'source' aDNA might be and how it might be read, let alone used, challenged, and narrated. If aDNA data is a 'source' then addressing and interrogating it would be an activity with an historiography and methodology. The notion of such work as 'technology' or 'tool' is common in the literature, providing a foundational sense that genetic analysis is purposeful, transitional, and of 'use' in historical investigation. The suggestion is that it creates a new archive of information that shifts our understanding of human history.

However some aDNA scholars go further than Rutherford's formulation. Ancient DNA can be argued to be a 'source' and 'tool' for historical investigation, something that can be addressed and interrogated by expert geneticist readers that we might, eventually, term 'historians'. Yet some of those who work on this material claim it to be more wide-ranging, arguing that the advent of advanced techniques leads to a category shift in the way that we know the past and the types of questions we might be able to ask and answer about it. For instance, Haak et al. and others suggest that analysis of aDNA presents an infrastructure for approaching the past. For them, genomic analysis is 'comparable in power to archaeology and linguistics' and has the capacity become an important disciplinary presence. It is a 'transformative technology' that brings knowledge and information.[82] It is poised to 'transform' our understanding of what 'history' is and how we can study it. Such writers imply that this is new historical practice rather than *simply* a tool or an extra dataset. It is way of knowing the past and enables a writing of that past in novel and innovative fashion.

For these groups aDNA work is not necessarily an approach but an epistemology, and this has implications for the way in which they envisage historical investigation. The scholars writing these papers and generating the information and making the assertions about the data are rarely professional historians. Research group studies of aDNA seek to write a 'genomic' history of the organism that refers to human organisation and society. They present the human body as a data archive. Whilst interacting with more recognisable accounts of the past, the work produced is something quite different, a biological and genetic outline of development and change across temporal moments. That is, it is a different *kind* of historical knowledge, with different purpose, evidence, and scope. The human subject here is the subspecies *Homo sapiens* rather than the individual within time. The imaginative, creative, and scientific tools that are being used to render this material conceptually and intellectually animate are those of the geneticist. Yet the reports are contributing to wider 'historical' understanding and challenging what 'history' is as a discipline.

This disciplinary challenge can be seen in David Reich's popular science book *Who We Are and How We Got Here,* a polemic arguing for the primacy of 'new' ways of investigating the past through DNA. The foundation of the book is the huge amount of work Reich has undertaken from his 'factory' laboratory in Harvard, added to the explosion in scholarship spearheaded by laboratories in Copenhagen, Beijing, London and Leipzig. Reich cites hundreds of new datasets from ancient individuals, suggesting that the ability to work on such a scale has led to a transformation in knowledge, arguing aDNA sequencing is a game-changer. Reich's rhetoric is excessive. It contrasts with careful analysis from some areas of the field, such as in within evolutionary anthropology, where the ethical implications of the work have long been uppermost in discussion and larger claims about historical impact are less important than careful assertions about population history.

However, the claims that aDNA work is changing perception of the human and human culture in ways that may concomitantly challenge long-held disciplinary and intellectual traditions is certainly the view of many geneticists. Leonardi et al. write 'Ancient DNA research has made massive progress in its rather short history', in particular moving into reclaiming extinct species 'greatly beyond the temporal range covered by museum specimens' (p. e1).[83] For these scholars investigation of aDNA is filling in gaps in knowledge of the past whilst contributing to understanding of the human in the present: 'the ability to recover whole genomes from ancient remains has emerged as a powerful tool for understanding the human past'.[84] As Fregel et al. argue, this has clear benefit for historical analysis: 'Over the last decade, paleogenomics has been instrumental in settling long-disputed archaeological questions'.[85]

Indeed many geneticists working in this area are highly vocal about its significance and impact. Iñigo Olalde and Carles Lalueza-Fox suggest 'We are currently entering in the golden era of paleogenomics, which will provide researchers with an unprecedented amount of information for the reconstruction of the human past'.[86] This explosion in data means that '*All* archaeological hypotheses involving population movements and affinities of cultural horizons will be explored from this emerging body of genomic data' (p. 7, my emphasis). Their vision is a huge archive of information that will provide the basis for all investigations of the past and the re-evaluation of existing positions. They point out that genomic

data is opening up 'past migratory movements, some of it *undetected* by archaeology' (my emphasis, p. 5).

David Reich argues that aDNA analysis is 'going to really profoundly change
the way we do archaeology, history, linguistics, sociology, even demography, and
even economic history' (my emphasis).[87] Iñigo Olalde and Carles Lalueza-Fox
suggest it necessitates a review of *all* historical knowledge. Joseph Pickrell and
David Reich argued in 2014:

> Because of these technological advances, the past few years have seen a
> dramatic increase in the *quantity* of data available for learning about human
> history. Equally important has been rapid innovation in methods for making
> inferences from these data. We argue here that the technological break-
> throughs of the past few years motivate a *systematic reevaluation of human
> history using modern genomic tools* – a new 'History and Geography of Human
> Genes' that exploits many *orders of magnitude* more data than the original
> synthesis. [my emphases][88]

They argue for a 'systematic reevaluation' of all of human history; this is predicated
upon more information and more data. Their argument here relates to the size and
high-definition of the information-archive. More information has been 'revealed',
so in their formulation more is *known*. In itself this presents historians with a
methodological challenge, as it is necessary to develop tools to read, understand,
and categorise this data. The rhetoric is revelatory ('breakthrough', 'modern',
dramatic', 'advance'). The revisionist movement that is suggested by the authors
here is to a certain extent characteristic of work in the field. New information
leads to new insight, the uncovering of a truth that was formerly obscured. The
information available for the study of the past has expanded. Such arguments
suggest that what is of prime importance is the gathering of data and the sophistication of analysis of this information. The historiographical contribution is to
augment what is known and to expand the archive. DNA data becomes the mass
of information to be addressed by the historian-geneticist. Their engagement with
this evidence necessitates an innovative set of methodological tools and a new way
of understanding what the 'past' actually *is*.

Interdisciplinarity and the Future of DNA-Led History

An article in the *New York Times Magazine* from 2019 presented some critical
positions and concerns about the new knowledge and the implications of aDNA
expansion:

> in practice, the paleogenomicists have totally altered the environment in
> which prehistory is being studied by everyone. The landscape is dominated
> by four well-funded, well-connected labs, three of which [...] collaborate

closely with one another, to the point that some critics accuse them of collusion. The power of these top labs extends to samples, data and even technology: Proprietary chemical reagents let them isolate and enrich ancient samples much more accurately and cost-effectively than other labs can.[89]

Many of the articles cited in this chapter reflect small but important shifts in the institutional and disciplinary location of pre-historical analysis. The influential scientific journals *Nature*, *Cell* and *Science* regularly publish on paleogenomics, paleoclimate, and genetic archaeology. Over the past two decades findings that relate to interpretation of the past have been presented in new scholarly locations. As interdisciplinary relationships between geneticists and humanities scholars develop further, increasingly research into the past will be undertaken within non-traditional areas and showcased in scientific journals. The large number of studies undertaken in the aDNA field in the past 15 years, and particularly since 2010, means that there is a huge amount of new data to engage with and understand.

Much aDNA practice engages with the disciplines of archaeology and anthropology.[90] The recent paleogenetic shift within prehistory studies, driven to an extent by the outpouring of data from a few influential laboratories, has led to a set of debates in archaeology about evidence, interdisciplinarity, and knowledge.[91] Anthropologists and scholars in critical race studies though have engaged with the ethics and implications of aDNA investigation, working through the 'consultation/collaboration/cooperation' options for those 'negotiating this space'.[92] The ethical and epistemological challenges of the new information have led to the development of new working models.[93] The swift expansion of aDNA work has led to a concomitant need for theoretical models and practices to be articulated. Scholars from a range of disciplines have attempted to articulate what type of approach aDNA-informed work might be, and how it might engage with existing structures and disciplinary approaches. A *Nature* editorial and news feature in 2018 outlined the 'several problems' surrounding the future of history, archaeology and aDNA multi-disciplinary study, particularly worrying about political interpretation of information; however *Nature* argued that 'presented correctly alongside insights from other disciplines, ancient-DNA research can be a powerful weapon against bigotry'.[94]

The status of aDNA knowledge within the discipline of archaeology is being debated fiercely, responding to the increasing numbers of papers being published by genetics research groups that impact upon the discipline.[95] There is a sophisticated response from archaeological theorists suggesting that, whilst the genetic work is impressive, the 'interpretational frameworks require more discussion'.[96] These responses have been concerned with the impact of aDNA on particular moments as scholars look to challenge the assumptions being made in the interpretation and presentation of the genetic data, agree terms, or interrogate the practices involved in the collection and interpretation of materials.[97] Many archaeologists are suspicious of what seems to be the return of grand narratives explaining swathes of human history. Similarly anthropologists have long discussed the interdisciplinary implications of aDNA work and reminded us that 'No aDNA study can proceed in isolation but is dependent on previous research [...] to provide a context'.[98] The nature of this context and the way that it is configured forces us to recognise aDNA research as a multi-disciplinary nexus.

Writing in the archaeological journal *Antiquity* a multidisciplinary group of authors argue that genetics has 'lifted an interpretative burden from archaeology'; the article is paired in the journal with a highly critical article by Volker Heyd.[99] The tone of the collaborative article is temperate, as the authors argue that they seek simply to 'reconstruct in some detail the social processes behind the observable archaeological changes' (p. 338). They argue for a plural approach to the materials of the past. The authors make revisionist claim and reach for revelatory vocabulary when outlining the implications of their work: 'Some may not like it for its resemblance to an older paradigm [...] but we are now in a position to *unravel* the *complexities* behind the historical processes in much more detail, and thus avoid the *simplistic* models of the past' (my emphases, p. 343). Kristiansen et al. in contrast envisage the future of historical investigation in terms of the team and collaborative aspect. The paper is authored by twelve scholars from a number of backgrounds and suggests future integration of scientific information into historical narrative and argument. The authors argue that their 'integrated model of cultural, linguistic and genetic change' challenges historical and archaeological assumptions and presents a new 'way' of approaching the data and combining work.

It might be that this kind of collaborative work will raise far more questions than it answers. Huw S. Groucutt et al. argue in an overview article:

> The emerging picture of the dispersal process suggests dynamic behavioural variability, complex interactions between populations, and intricate genetic and cultural legacy. This evolutionary and historical complexity challenges simple narratives and suggests that hybrid models and the testing of explicit hypotheses are required[100]

The call for 'hybrid models' suggests a particular type of interdisciplinary process is needed to replace 'simple narratives'. The complex interaction of a variety of disciplinary models might not produce clarity. The evidence is seemingly contradictory: 'fossil, genetics, and archaeological data are currently consistent with several different models' (p. 161). The amount of data that is being generated is extraordinary. Indeed, the multiplicity of data and information, as much as the diversity of approach, might lead to a *lack* of clarity.

Kristiansen et al.'s invocation of an 'integrated' investigative model, and Groucutt et al.'s sense of 'hybrid', both require further parsing in terms of what they imply for interdisciplinarity, multi-disciplinarity, and the methodological logistics of future investigation.[101] The assimilative-integrative concept itself has a prehistory, as the term 'integrated' responds to Colin Renfrew's repeated calls for a synthesis 'between genetics, archaeology and linguistics'.[102] It reflects Cavalli-Sforza's hope that archaeology, linguistics and genetics will 'converge' toward 'a common story [...] a single history'.[103] The assimilative model suggests that aDNA work will substantially contribute to investigation, taking its place as one of the three key approaches to prehistory. The terminologies ('integrated'; 'hybrid') themselves use a vocabulary of admixture and integration that should be critiqued. The concept of a mixed approach, a synthesis, or a yoking of the three approaches demands careful consideration of what will happen in practice and what this implies about the structure of knowledge being propounded. It is uncertain exactly how 'integrated' structures would work, given the current lack of methodological engagement between geneticists and humanities scholars. If the model is simply to package out sections of investigation to various disciplines, then this does not equate to an interdisciplinary practice. However, it seems certain that historians will need to develop ways of responding to genetic data and the ability to understand the implications of DNA information.

Scholars have begun to engage with and use the new data yielded by and new approaches enabled by aDNA analysis. Yet, as one critique puts it, the 'the frequent lack of genuine collaboration between fields' has slowed the process.[104] Brooke and Larsen call for historians to 'engage closely' with geneticists: 'Much of this terrain is new, and perhaps forbidding to historians trained to think about documents and decades, but we need to take up the challenge – and to have an active role in this work' (p. 1513). Polemics and perspectives like this seek interdisciplinary traction and encourage historians to engage with the new areas as much as possible, to be contributory partners in developing a new tool for the investigation of the past. One prominent commentator recently recognised 'tensions' but emphasising the 'opportunity' for a 'transdisciplinary approach' and the importance of profound collaboration in practice in a mode is yet to be defined.[105] Yet as has been seen, many research groups are making their historical-revisionist case already, albeit without any kind of formalised method. Research groups are reading the data with polemic and historiographic purpose, creating an historicised genomics that has impact upon wider public understanding of the past.[106] It also has a influence on the type of aDNA research undertaken. The types of approach that are being considered and articulated demonstrate the volatility of the intellectual situation.

Should historians happily acclaim the arrival of 'a new player onto the field of historical research' or remain more sceptical?[107] The aDNA genii cannot be put back in the bottle, and historians must ponder how to assimilate such methodologies within current historical frameworks or possibly evolve new approaches.[108] Keith Wailoo, Alondra Nelson and Catherine Lee argue that humanities scholars need to critique DNA work of all kinds:

> [DNA analysis] makes fundamental, if problematic, claims about the present and the distant past – and as such, the claims, credibility, and applications of the genetic science must be examined closely and in multiple venues.[109]

Whilst the formal historiography of such aDNA investigation has not yet been written, and practice has evolved quickly and organically, significant investigations are being undertaken by research teams and seemingly impressive results published. At present the orthodoxy and approach of aDNA work is being established through scientific practice and dictated by major global research networks.[110] For research groupings around the world aDNA analysis is a tool that enables huge change in the *way* that the past is studied and indeed *what* is studied. The challenge is to understand the new models and innovative languages available for investigating, understanding, and 'telling' the past.[111] Disaggregating excessive rhetoric from epistemic change, and interrogating 'new' modes of knowledge that present themselves as radical and transformative, is the task of the historian as it already has been for the archaeologist and the anthropologist. It is crucial to remain sceptical of the idea of a transformation in knowledge and to resist some of the more totalising claims being made whilst seeking to recognise the genuine potential for an innovative bio-history.[112]

Ancient DNA presents use with seemingly contradictory outcomes. This type of work challenges an anthropocentrism in the study of the past that chimes with other work in the academy. In particular, aDNA work seems to give agency to non-human actors and objects, a prehistoric posthumanism that conceives of *Homo sapiens* as information rather than culture. Research into pathogens and non-human aDNA suggests an engagement with non-human actions in history. The scope and scale of much of the work clearly has much to contribute to debates within 'big' history and studies of 'deep time'.[113] At the same time work on *Homo sapiens* aDNA places the 'human' centrally, albeit in a way that renders in flux what exactly the 'human' is. This complication of the human under consideration suggests that aDNA work has a liberating context for approaches to the past. Conversely it might demonstrate an essentialism about particular ways of thinking about the human, as has been provoked by discussions of race and ethnicity in relation to such investigation. DNA work has sometimes been undertaken in close contact with indigenous groups.[114] This has had some effect in raising awareness of issues relating to commemoration, land rights, and historical marginalisation. However, the discussions of ethnic identity found in genetics papers and popular work have attracted controversy particularly due to assumptions about how to categorise and identify race.[115] Ancient DNA work challenges anthropocentric assumptions of 'history' as an approach and set of institutions.[116] At the same time aDNA works to situate a 'modern' human and an 'ancient' set of remains and data.

Addressing this complexity demonstrates the ambiguity inherent in this set of approaches. We might think of aDNA investigation as a kind of memorial practice, a contemporary mode of remembering and knowing the past, of reconstructing, re-enacting, and mourning. We can further suggest it has ideological affinity with particularly neo-liberal knowledge organisation, and that often it seems to reassert a kind of normativity and problematic essentialism. We could situate the new postgenomic age as a moment in which genetic discourse has begun to engulf all of human life, past and present. The opening of an aDNA laboratory at the Francis Crick Biomedical Discovery Institute in London in December 2017, for instance, makes evident the institutional links between such investigation and contemporary health research. We might dismiss aDNA work as irrelevant, claim it to be worthless or pointless. Or we might argue it offers a means of glimpsing a different type of past.

Given the size of the datasets now being generated, and the institutional support now being offered this work, aDNA will be difficult to ignore in the coming decade. The interrogation of central issues here pertaining to evidence, memory, the manipulation of data, the storage of information, discussion of human migration, the internalisation of complexity, the development of historiography through practice, and, above all, the ethics of historical inquiry, needs to be undertaken by historians of all kinds with some expedience.[117] David Reich argues that 'I think what the DNA is doing is it is forcing the hand of this discussion in archaeology'.[118] The scope of this forceful quality is challenging discussions across the historical spectrum. Historians must interrogate the production and presentation of this genetic evidence and data, to scrutinise the practice of this work and submit it to robust critique. History as a discipline needs to be able to recognise and understand the contribution of geneticists, and to consider whether their analysis of ancient samples is a kind of historical practice. If it is, we need to investigate *how* it can be understood as such and what the implications are of this for the discipline.

Notes

1 See Svante Pääbo et al., 'Genetic Analyses from Ancient DNA', *Annual Review of Genetics* xxxviii (2004), 645–79, Eske Willerslev and Alan Cooper, 'Review Paper: Ancient DNA', *Proceedings of the Royal Society B* cclxxii (2005), 3–16, and Nicolas Arning and Daniel J. Wilson, 'The past, present and future of ancient bacterial DNA', *Microbial Genomics* 6:7 (2020), doi: 10.1099/mgen.0.000384. Thanks to Tom Booth for suggestions on this latter point.

2 Reich, *Who we are and how we got here*, p. xviii, Kristiansen, 'Towards a New Paradigm?', *Current Swedish Archaeology*, 22 (2014), 11–34.

3 Ewen Callaway, 'Divided by DNA', *Nature* 555 (2018), 573–76.

4 Ewen Callaway, 'Ancient DNA reveals secrets of human history', *Nature* 476 (2011), 136–7 (136).

5 Wolfgang Haak et al., 'Massive migration from the steppe was a source for Indo-European languages, *Nature* dxxii (2015), 207–211 (207). The progressive ideal of a 'transformative technology' has been well critiqued by for instance Andrew Feenberg, *Transformative Technology: A Critical Theory Revisited* (Oxford: Oxford University Press, 2002).

6 Build monica green and Lester Little footnote https://academic.oup.com/shm/article-abstract/23/3/701/1720927.

7 There is an account of some of the general issues for DNA and history in John L. Brooke and Clark Spencer Larsen, 'The Nurture of Nature: Genetics, Epigenetics, and Environment in Human Biohistory', *The American Historical Review* 119: 5 (2014) 1500–1513.

8 See Terence D. Keel, 'Human-Neanderthal Hybrids and the Frontier of Critical Race Studies' in *Red and Yellow, Black and Brown: Decentering Whiteness in Mixed Race Studies* ed. Paul Spickard and Rudy Guevarra (New Brunswick, NJ: Rutgers University Press, 2017), pp. 201–18.

9 For a discussion of the development of aDNA sequencing techniques see Michael Hofreiter, David Serre, Hendrik N. Poinar, Melanie Kuch, and Svante Pääbo, 'Ancient DNA', *Nature Reviews: Genetics* 2 (2001), 353–59. Pääbo also outlines the development of the technology in his book *Neanderthal Man* (London: Basic Books, 2015).

10 Svante Pääbo, 'Preservation of DNA in ancient Egyptian mummies', Journal of Archaeological Science 12:6, 411–17.

11 See Venla Oikkonen, *Population Genetics and Belonging* (Basingstoke: Palgrave, 2018), pp. 79–105.

12 Erika Hagelberg, Bryan Sykes, and Robert Hedges, 'Ancient bone DNA amplified', *Nature* 342: 485 (1989), 485.

13 Frederika A. Kaestle and K. Ann Horsburgh, 'Ancient DNA in Anthropology: Methods, Applications, and Ethics', *Yearbook of Physical Anthropology* xlv (2002), 92–130 and Dennis H. O'Rourke, M. Geoffrey Hayes, Shawn W. Carlyle, 'Ancient DNA Studies in Physical Anthropology', *Annual Review of Anthropology* 29 (2000), 217–42.

14 'Ancient DNA and the Archaeologist', *Antiquity* 66: 250 (1992), 10–14.

15 Kaestle and Horsburgh, 'Ancient DNA', pp. 93, 109.

16 Casey C. Bennett and Frederika A. Kaestle, 'Reanalysis of Eurasian Population History: Ancient DNA Evidence of Population Affinities', *Human Biology* 78:4 (2006), 413–440.

17 Oikkonen, *Population Genetics and Belonging*, p. 81.

18 Alan Cooper, 'Ancient DNA: do it right or not at all', *Science* 289:5482 (2000), p. 1139–41.

19 The first aDNA 'next generation' work included M. Margulies et al., 'Genome sequencing in microfabricated high-density picolitre reactors', *Nature* 437 (2005), 376–380. For an overview of the impact of 'next generation' technology see Michael Knapp and Michael Hofreiter, 'Next Generation Sequencing of Ancient DNA: Requirements, Strategies and Perspectives', *Genes* 1:2 (2010), 227–43.

20 Andreas Keller et al., 'New insights into the Tyrolean Iceman's origin and phenotype as inferred by whole-genome sequencing', *Nature Communications* 3 article 698 (2012), https://doi.org/10.1038/ncomms1701.

21 R.E. Green et al., 'A Draft Sequence of the Neandertal Genome', *Science* 328 (2010), 710–22, David Reich et al., 'Genetic History of an Archaic Hominin Group from Denisova Cave in Siberia', *Nature* 468 (2010) and M. Rasmussen et al., 'Ancient Human Genome Sequence of an Extinct Paleo-Eskimo', *Nature* 463 (2010), 757–62.

22 Marc Haber, Massimo Mezzavilla, Yali Xue, and Chris Tyler-Smith, 'Ancient DNA and the rewriting of human history', *Genome Biology* xvii (2016), doi: 10.1186/s13059-015-0866-z. See the critical account of how aDNA work has become focused on a few labs telling 'heroic' stories in Erika Hagelberg, Challenging the Narratives in Ancient DNA, in Kallen *Game of Bones* ref needed.

23 Mathieson and Skoglund, 'Ancient Human Genomics: the first decade', *Annual Review of Genomics and Human Genetics* xix (2018), https://doi.org/10.1146/annurev-genom-083117-021749.

24 Callaway, 'Ancient DNA reveals secrets of human history', 136.

25 Roseina Woods, Melissa M. Marr, Selina Brace, and Ian Barnes, 'The Small and the Dead: A review of Ancient DNA Studies Analysing Micromammal Species', *Genes* 8:312 (2017), doi:10.3390/genes8110312.

26 Callaway, 'Ancient DNA reveals secrets of human history', 136. On the language of excess in aDNA work see Elizabeth D. Jones and Elsbeth Bösl, 'Ancient human DNA: A history of hype (now and then), *Journal of Social Archaeology* (2021), https://doi.org/10.1177/1469605321990115.

27 Michela Leonardi et al., 'Evolutionary Patterns and Processes: Lessons from Ancient DNA', *Systematic Biology* lxvi (2017), e1–e29 (p. e1).

28 Stephanie Marciniak and George H. Perry, 'Harnessing ancient genomes to study the history of human adaptation', *Nature Reviews Genetics* 18 (2017), 659–74.

29 'BP' means 'Before Present' and is a convention of radiocarbon dating that generally takes 1950 as 'present', E.W. Wolff, 'When is the "present"?', *Quaternary Science Reviews* 27 (2007), 3023–4.

30 Michael Bunce et al., 'Ancient DNA provides new insights into the evolutionary history of New Zealand's extinct giant eagle', *PLoS Biology* 3:1 (2005), https://doi.org/10.1371/journal.pbio.0030009.

31 Beth Shapiro, *How to Clone a Mammoth: The Science of De-exctinction* (Princeton, NJ: Princeton University Press, 2015), p. 55.

32 Maria A. Nieves-Colón et al., 'Comparison of two ancient DNA extraction protocols for skeletal remains from tropical environments', *American Journal of Physical Anthropology* 166 (2018), 824–36. On the increased ability to gather aDNA in Central Africa see Ann Gibbons, 'DNA from child burials reveals 'profoundly different' human landscape in ancient Africa', *Science* 22 January 2020, https://www.sciencemag.org/news/2020/01/dna-child-burials-reveals-profoundly-different-human-landscape-ancient-africa?utm_campaign=ScienceNow&utm_source=Contractor&utm_medium=Facebook [accessed 13 May 2021].

33 Iñigo Olalde et al., 'The Beaker phenomenon and the genomic transformation of northwest Europe', *Nature* dlv (2018), doi:10.1038/nature25738.

34 Cristina Gamba et al., 'Genome flux and stasis in a five millennium transect of European prehistory', *Nature Communications* v (2014), 1–9 (1); Qiaomei Fu et al., 'An early modern human from Romania with a recent Neanderthal ancestor', *Nature* dxxiv (2015), 216–31.

35 Chris Clarkson et al., 'Human occupation of northern Australia by 65000 years ago', *Nature* dxlvii (2017), 306–26.

36 Qiaomei Fu et al., 'An early modern human from Romania with a recent Neanderthal ancestor', *Nature* dxxiv (2015), 216–31; Douglas J. Kennett et al., 'Archaeogenetic evidence reveals prehistoric matrilineal dynasty', *Nature Communications* vii (2017), doi: https://doi.org/10.1038/ncomms14115.

37 Diana I. Cruz-Dávalos et al., 'In-solution Y-chromosome capture-enrichment on ancient DNA libraries', *BMC Genomics* 19: 608 (2018), https://doi.org/10.1186/s12864-018-4945-x.

38 Morten Rasmussen et al., 'The genome of a Late Pleistocene human from a Clovis burial site', *Nature* dvi (2014), 225–9.

39 Éadaoin Harney et al., 'Ancient DNA from the skeletons of Roopkund lake reveals Mediterranean migrants in India', *Nature Communications* 10 article 3670 (2019), https://doi.org/10.1038/s41467-019-11357-9.

40 Joannella Morales et al., 'A standardized framework for representation of ancestry data in genomics studies', *Genome Biology* ix (2018), https://doi.org/10.1186/s13059-018-1396-2, Alice B. Popejoy and Stephanie M. Fulleton, 'Genomics is failing on diversity', *Nature* dxxxviii (2016), 161–5l, Jessica Bardill et al., 'Advancing the ethics of paleogenomics', *Science*, ccclx (2018), 384–5.

41 On modern aDNA work see for instance Ivan Jerković, Željana Bašić, Ivana Kružić, and Šimun Anđelinović, 'Creating reference data on sex for ancient populations using the Probabilistic Sex Diagnosis method', *Journal of Archaeological Science* xciv (2018), 44–50, Ben Krause-Kyora et al., 'Ancient DNA study reveals HLA susceptibility locus for leprosy in medieval Europeans', *Nature Communications* ix (2018), https://doi.org/10.1038/s41467-018-03857-x.

42 Hannes Schroeder et al., 'Genome-wide ancestry of 17th-century enslaved Africans from the Caribbean', *PNAS* 112 (2015), 3669–73; Niall O'Sullivan et al., 'Ancient genome-wide analyses infer kinship structure in an Early Medieval Alemannic graveyard', *Science Advances* iv (2018), doi: https://doi.org/10.1126/sciadv.aao1262; Carlos Eduardo G. Amorim et al., 'Understanding 6th-century barbarian social organization and migration through paleogenomics', *Nature Communications* ix (2018), doi: http://doi.org/10.1038/s41467-018-06024-4.

43 O'Sullivan et al; Schroeder et al.

44 Jada Benn Torres, 'Genetic Anthropology and Archaeology: Interdisciplinary Approaches to Human History in the Caribbean', *PaleoAmerica* 2:1 (2016), 1–5 (p. 4).

45 Richard Buckley, Mathew Morris, Jo Appleby, Turi King, Deirdre O'Sullivan and Lin Foxhall, "The king in the car park": new light on the death and burial of Richard III in the Grey Friars church, Leicester, in 1485', *Antiquity*, lxxxvii (2013), 519–538.

46 See Oikkonen, *Population Genetics and Belonging*, pp. 92–105.

47 Ewen Callaway, 'Ancient genome stirs ethics debate', *Nature* dvi (2014), 142–3.

48 Karin Margarita Frei et al., 'Tracing the dynamic life story of a Bronze Age Female', *Nature Scientific Reports* 5 article 10431 (2015), https://doi.org/10.1038/srep10431.

49 Carl Zimmer, 'Eske Willerslev is Rewriting History with DNA', *New York Times*, 16 May 2016, https://goo.gl/r5hpm8 [accessed 16 August 2018].

50 Paul Rincon, 'How ancient DNA is transforming our view of the past', 12 April 2018, https://goo.gl/Zwkbzm [accessed 16 August 2018].

51 Sarah Zhang, 'Ancient DNA is Rewriting Human (and Neanderthal) History', *The Atlantic*, 14 March 2018, https://goo.gl/LEVsEv [accessed 27 March 2018]; Sarah Kaplan, 'Neanderthal microbes', 8 March 2017, https://goo.gl/ocB5EU [accessed 16 August 2018]; Cheyenne Macdonald, 'Groundbreaking study', *Daily Mail*, 10 July 2018, https://goo.gl/bJtbtB [accessed 16 August 2018].

52 Robin KcKie, 'Tiny traces of DNA found in cave dust', *The Observer*, 16 May 2021, https://www.theguardian.com/science/2021/may/16/tiny-traces-of-dna-found-in-cave-dust-may-unlock-secret-life-of-neanderthals [accessed 19 May 2021].

53 Editorial team, 'Ancient DNA cracks puzzle of Basque origins', *BBC*, 7 September 2015, https://www.bbc.co.uk/news/science-environment-34175224 [accessed 14 May 2021].

54 Torsten Günther et al., 'Ancient genomes link early farmers from Atapuerca in Spain to modern-day Basques', *PNAS* 112: 38 (2015), 11917–11922.

55 *Ancestors* (London: Simon & Schuster, 2021), Kindle edition paragraph 15.

56 Recent overviews of aDNA technical methodologies include Bolnick et al., 'Native American Genomics' and Joannella Morales et al., 'A standardized framework for representation of ancestry data in genomics studies, with application to the NHGRI-EBI GWAS Catalog', *Genome Biology* 19:21 (2018), https://doi.org/10.1186/s13059-018-1396-2. See also the handbook *Ancient DNA: Methods and Protocols* ed. Beth Shapiro, Axel Barlow, Peter D. Heintzman, Michael Hofreiter, Johanna L.A. Paijmans and André E.R. Soares (Basingstoke: Springer Nature, 2019).

57 See Iain Mathieson et al., 'The Genomic History of Southeastern Europe', *Nature* dlv (2018), 197-203 and Anna-Sapfo Malaspinas et al., 'A genomic history of Aboriginal Australia', *Nature* dxxxviii (2016), 207–14.

58 Morten Rasmussen et al., 'The genome of a Late Pleistocene human from a Clovis burial site', *Nature* dvi (2014), 225-9; Anuradha Jagadeesan et al., 'Reconstructing an African haploid genome from the 18th century', *Nature Genetics* l (2018), 199–205; Hideaki Kanzawa-Kiriyama et al., 'A partial nuclear genome of the Jomons who lived 3000 years ago in Fukishima', *Journal of Human Genetics* lxii (2017), 213–221.

59 Fu et al., 'The genetic history of Ice Age Europe', *Nature* dxxxiv (2016), 200–5.

60 on contamination see Pontus Skoglund et al., 'Separating endogenous ancient DNA from modern day contamination', *PNAS* cxi (2014), 2229–34 (2230) and Bastien Llamas et al., 'Controlling DNA contamination in human ancient DNA research in the high-throughput sequencing era', *STAR* iii (2017), 1–14.

61 Peter de Barros Damgaard et al., '137 ancient human genomes from across the Eurasian steppes', *Nature* dlvii (2018), 369–74.

62 Mathieson et al., 'The genomic history of southeastern Europe', 197; Anna-Sapfo Malaspinas et al., 'A genomic history of Aboriginal Australia'.

63 Lazaridis et al., 'Genomic insights into the origin of farming in the ancient Near East', *Nature* dxxxvi (2016), 419–26.

64 Kristiansen et al., 'Re-theorising mobility and the formation of culture and language among the Corded Ware Culture in Europe', *Antiquity* xci (2017), 334–47 (343).

65 Fregel et al., 'Ancient genomes from North Africa', p. 6774.

66 *Le Monde*, 29 August 2006, https://goo.gl/5rWeuB [accessed 16 August 2018].

67 Gideon Lewis-Kraus, 'Is Ancient DNA research revealing new truths – or falling into old traps?', *New York Times*, 17 January 2019, https://www.nytimes.com/2019/01/17/magazine/ancient-dna-paleogenomics.html [accessed 19 June 2019].

68 Patcharee Lertit et al., 'Genetic History of Southeast Populations as Revealed by Ancient and Modern Human Mitochondrial DNA Analysis', *American Journal of Physical Anthropology* 137 (2008), 425–40.

69 Joseph K. Pickrell et al., 'The genetic prehistory of southern Africa', *Nature Communications* iii (2012), doi: 10/1038/ncomms2140.

70 Alissa Mittnik et al., 'The genetic prehistory of the Baltic Sea region', *Nature Communications* ix (2018), DOI: 10.1038/s41467-018-02825.

71 Tara L. Fulton and Beth Shapiro, 'Setting up an Ancient DNA lab' in Shaprio et al., eds., *Ancient DNA*, pp. 1–13 (p. 1).

72 Marc Vander-Linden, 'Toward a clearer view into human prehistory', *Science* 363: 6432 (2019), 1153–4.

73 Hannah Devlin, 'First modern Britons had 'dark to black' skin', *Guardian*, 7 February 2018, https://goo.gl/ouA5dJ and David Keys, 'Britain's prehistoric catastrophe revealed', *Independent*, 21 February 2018, https://goo.gl/d7YdfL [accessed 30 October 2018].

74 There has always been of course much discussion about the term 'Ancient DNA', see, for instance, D.H. O'Rourke, S.W. Carlyle, and R. L. Parr, 'Ancient DNA: Methods, Progress, and Perspectives', *American Journal of Human Biology* 8 (1996), 557–71.

75 on prehistory see Donald R. Kelley, 'The Rise of Prehistory', *Journal of World History* xiv (2003), 1–19.

76 Rasmussen et al., 'Ancient Human Genome Sequence of an Extinct Palaeo-Eskimo', Reich et al., 'Genetic History of an Archaic Hominim Group from Denisova Cave in Siberia', Green et al., 'A Draft Sequence of the Neanderthal Genome'.

77 Maarten Blaauw et al., 'The Problems of Radiocarbon Dating', *Science* cccvii (2005), 1551–3.

78 Iosif Lazaridis et al., 'Genomic insights into the origin of farming in the ancient Near East', Lars Fehren-Schmitz et al., 'Genetic Ancestry of Rapanui before and after European Contact', Andain Seguin-Orlando et al., 'Genomic structure in Europeans dating back at least 36,200 years', *Science* cccxlvi (2014), 1113-18 (1113), Iñigo Olalde et al., 'The Beaker phenomenon and the genomic transformation of northwest Europe'.

79 On periodisation, see Daniel Lord Smail and Andrew Shryock, 'History and the "Pre"', *American Historical Review* cxviii (2013), 709–737.

80 Clark Spencer Larsen, *Bioarchaeology* (Cambridge: Cambridge University Press, 2015, first published 1997).

81 *A Brief History of Everyone Who Ever Lived* (London: 2017), p. 4.

82 Haak et al., 'Massive migration', p. 207.

83 Michela Leonardi et al., 'Evolutionary Patterns and Processes: Lessons from Ancient DNA', p. e1; Sriram Sankararaman et al., 'The genomic landscape of Neanderthal ancestry in present-day humans', *Nature*, dvii (2014), 354–7.

84 Niels N. Johanssen, Greger Larson, David J. Metzler, and Marc Vander Linden, 'A composite window into human history', *Science* ccclvi (2017), 1118–20 (1118).

85 Rosa Fregel et al., 'Ancient genomes from North Africa evidence prehistoric migrations to the Maghreb from both the Levant and Europe', *PNAS* 115:26 (2018), 6774–9 (p. 6774).

86 'Modern humans' paleogenomics and the new evidences on the European prehistory', *STAR: Science & Technology of Archaeological Research* i (2015), 1–9 (7).

87 https://www.pbs.org/newshour/show/the-ancient-dna-revolution-unlocks-how-connected-we-all-are [accessed 14 September 2018].

88 Joseph Pickrell and David Reich, 'Toward a new history and geography of human genes', *Trends in Genetics*, xxx (2014), 377–89 (378). The interrelationship of genes and geography is criticised in Jenny Reardon, *Race to the Finish* (Princeton, NJ: 2005).

89 Gideon Lewis-Kraus, 'Is Ancient DNA Research Revealing New Truths – or Falling into Old Traps?', *New York Times Magazine* 17 January 2019, https://www.nytimes.com/2019/01/17/magazine/ancient-dna-paleogenomics.html [accessed 19 May 2021].

90 For anthropological discussion of aDNA see Michael Banton, 'Genomics and Race: Vexed Questions', *Anthropology Today* 21:4 (2005), 3–4.

91 Marc Vander Linden, 'Population history in third-millennium-BC Europe: assessing the contribution of genetics', *World Archaeology* xlviii (2016), 714–28 (714), Volker Heyd, 'Kossinna's smile', *Antiquity* xci (2017), 348–59 (p. 354).

92 Kaestle and Horsburgh, 'Ancient DNA in Anthropology', p. 108. See also John F. Hoffecker, Scott A. Elias, Dennis H. O'Rourke, G. Richard Scott and Nancy H. Bigelow, 'Beringia and the global dispersal of modern humans', *Evolutionary Anthropology* 25 (2016), 64–78 and Deborah A. Bolnick, Jennifer A. Raff, Lauren C. Springs, Austin W. Reynolds, and Aida T. Miró-Herrans, 'Native American Genomics and Population Histories', *Annual Review of Anthropology* 45 (2016), 319–40.

93 See for instance Stefanie Eisenmann et al., 'Reconciling material clusters in archaeology with genetic data: the nomenclature of clusters emerging from archeogenomic analysis', *Scientific Reports* 8 (2018), doi: 10.1038/s41598-018-31123-z.

94 Editorial 'On the use and abuse of Ancient DNA', *Nature* 555 (2018), 559 and Ewen Callaway, 'The Battle for Common Ground', *Nature* 555 (2018), 573–6.

95 Joanna Brück, 'Ancient DNA, kinship and relational identities in Bronze Age Britain', *Antiquity* 35:379 (2021), 228–37.

96 Furholt, 'Massive Migrations?', p. 159; some archaeology responses include Daniela Hofmann, 'What Have Genetics Ever Done for Us?: Implications of aDNA Data for Interpreting Identity in Early Neolithic Central Europe', *European Journal of Archaeology* xviii (2015), 454–76 and Mary E. Prendergast and Elizabeth Sawchuk, 'Boots on the ground in Africa's ancient DNA 'revolution': archaeological perspectives on ethics and best practices', *Antiquity* xcii (2018), 803–15.

97 Michael Banton, 'Genomics and Race: Vexed Questions', *Anthropology Today* 21:4 (2005), 3–4.

98 Connie J. Mulligan, 'Anthropological Applications of Ancient DNA: Problems and Prospects', *American Antiquity* 71:2 (2006), 365–80.

99 Kristian Kristiansen et al., 'Re-theorising mobility', 335; Heyd, 'Kossinna's smile', p. 354.

100 Huw S. Groucutt et al., 'Rethinking the Dispersal of *Homo sapiens* out of Africa', *Evolutionary Anthropology* xxiv (2015), 149–64 (149).

101 See Alexandra Ion, 'How Interdisciplinary is Interdisciplinarity? Revisiting the Impact of aDNA Research for the Archaeology of Human Remains', *Current Swedish Archaeology* xxv (2017), 177–98.

102 Colin Renfrew, 'Archaeogenetics', *Current Biology*, xx (2010), 162–165 (162).

103 *Genes, Peoples and Languages* (New York, NY: 2000), p. vii. See the critical discussion of this in Reardon, *Race to the Finish*, and Sommer, *History Within*. See Duanna Fullwiley, 'The "Contemporary Synthesis": When Politically Inclusive Genomic Science Relies on Biological Notions of Race', *Isis* cv (2014), 803–14.

104 Johanssen, Larson, Metzler, and Vander Linden, 'A composite window into human history', 1119.

105 Vander Linden, 'Towards a clearer understanding', p. 1154.

106 Alissa Mittnik et al., 'Kinship-based social inequality in Bronze Age Europe', *Science* 366 (2019), 731–4.

107 Monica H. Green, 'Genetics as a historicist discipline: a new player in disease history', *Perspectives on History*, December 2014, https://goo.gl/rDViQf; for an example of how critical historiography might engage with paleogenomics see Terrence Keel, *Divine Variations* (Stanford, CA: Stanford University Press, 2018).

108 See the special edition of essays on 'Pandemic Disease in the Medieval World: Rethinking the Black Death' edited by Monica H. Green and Carol Symes for *The Medieval Globe* i (2015), https://scholarworks.wmich.edu/tmg/vol1/iss1/1/, and Lester K. Little, 'Plague Historians in Lab Coats', *Past and Present*, ccxiii (2011), 267–90.

109 'Introduction' in Keith Wailoo, Alondra Nelson and Catherine Lee, eds., *Genetics and the Unsettled Past* (New Brunswick, NJ: Rutgers University Press, 2012), pp. 1–12 (p. 7).

110 the archaeological critique of aDNA is beginning to develop, see Martin Furholt, 'Massive Migrations? The Impact of Recent aDNA Studies on our View of Third Millennium Europe', *European Journal of Archaeology* xxi (2018), 159–91.

111 Critical voices from within Genetic Anthropology include for instance the Paleogenomics lab at UCLA, see Rosa Fregel et al., 'Ancient genomes from North Africa evidence prehistoric migrations to the Maghreb from both the Levant and Europe', *Proceedings of the National Academy of Sciences* 115:26 (2018), 6774–9.

112 Critical work on DNA more generally has come from the social sciences in particular Science and Technology Studies and Anthropology, see Alondra Nelson, *The Social Life of DNA* (Boston, MA: 2016), Catherine Nash, *Genetic Geographies* (Minneapolis, MN: 2016), Donna Harraway, *Modest_Witness* (London and New York: Routledge, 1997), Kim TallBear, *Native American DNA* (Minneapolis, MN: 2013); the classic critique is Dorothy Nelkin and M. Susan Lindee, *The DNA Mystique: the gene as cultural icon* (New York, NY: 1995).

113 David Christian, *Maps of Time: an Introduction to Big History* (Berkeley, CA: University of California Press, 2004).

114 Jennifer A. Raff and Deborah A. Bolnick, 'Paleogenenomics: Genetic roots of the first Americans', *Nature* dvi (2014), 162-3.

115 Daryl Leroux, '"We've been here for 2,000 years": White settlers, Native American DNA and the phenomenon of indigenization', *Social Studies of Science* xlviii (2018), 80–100.

116 Joan H. Fijimura and Ramya Rajagopalan, 'Different differences: The use of 'genetic ancestry' versus race in biomedical human genetic research', *Social Studies of Science* xli (2011), 5–30.

117 See Thomas Laqueur, *The Work of the Dead* (Princeton, NJ: 2015) and Crandall and Martin, 'The Bioarchaeology of Postmortem agency', *Cambridge Archaeological Journal* xxiv (2014), 429–35.

118 Zhang, 'Ancient DNA is Rewriting Human (and Neanderthal) History'.

3
POLITICS

The suggestion by a team of geneticists and archaeologists that the 10000 year-old 'Cheddar Man', also known as the 'First Brit', had dark skin pigmentation led to a wide-ranging political discussion about identity and nationhood in 2018.[1] This chapter considers discussions of genetics and race to confirm that postgenomic science is not politically neutral. As explored in the previous chapter, the collection, interrogation, interpretation and arrangement of genetic information reflects structures of power and invokes an historiographical ethical dimension which is often underexamined.[2] As evidence DNA information must be interrogated in multiple ways, and as a part of an historical narrative it needs to be considered in as wide a context as possible. Yet clearly genetic data is not neutral in its collection, storage, usage and interpretation. This chapter reviews high-profile debates over ethnicity and identity as related to genetic investigation of the past. It also analyses conflicted genetic histories by looking at recent cases where local, domestic, and Indigenous groups have challenged the findings and science of genomics. I look at the ways in which prehistory has been conceptualised as 'before' race and explore this in relation to discussions of nationalism in the United Kingdom. To conclude I present ways in which genetic data and DNA-led models of ethnicity and history could be challenged and interrogated.

Biocolonialism and Political Power

The opening up of new archives of information and genomic data has huge potential to 'empower Indigenous communities' by 'repatriating our deep past'.[3] For multiple groups DNA might enable connection and ownership of pasts that have often been violently erased or stolen. However, many of the contemporary ways in which genetic data is gathered, managed, presented, and interpreted often 'only reaffirms Indigenous peoples' fears of biocolonialism'.[4] This section illustrates

DOI: 10.4324/9781003052975-4

some of the highly problematic ways that genetic science has dealt with Indigenous and native communities to flag some of the central problems in considering this approach to knowing and articulating the past.

The Human Genome Diversity Project

The Human Genome Project did not consider human ethnic variation. In 1991 those interested in population genetics suggested an additional Human Genome Diversity Project, claiming that the HGP had 'a vanishing opportunity to preserve the record of our genetic heritage'.[5] They argued that the 'populations that can tell us most about our evolutionary past are those that have been isolated for some time', adding that 'Isolated human populations contain much more informative genetic records than more recent, urban ones' (p. 490). Their intervention was historical as well as scientific, as they suggested: 'We must act now to preserve our common heritage' and to save 'this historic record' (p. 490). The purpose of the exercise was archival, insofar as it sought to preserve material for future work to be undertaken. It is important to note here the confluence of humanist ideals of understanding the species with medical necessity, as the impetus to collect genetic information was 'to understand human diversity, both normal variation and that responsible for inherited diseases' (p. 490). The proponents of the HGDP mention specifically peoples from the Sahara, Japan, Malaysia, China, Polynesia, as well as Aboriginal and 'Indigenous American populations' (p. 490), suggesting that 'It will be essential to integrate the study of peoples with response to their related needs' (p. 490).

The 'historical' claim that genetics is common human heritage compares DNA information to heritage materials such as might be found in a museum, an unfortunate linkage given the struggles of Indigenous and native peoples to have their artefacts returned or to resist being 'categorised' by well-meaning scientists, curators, and anthropologists.[6] As Catherine Nash suggests, the HGDP utilised a problematic 'rhetoric of genetic salvage', and this sense of needing to preserve genetic information from the past for some kind of use in the present and future led to many groups challenging or ignoring the initiative.[7] Indeed from inception HGDP was criticised by Indigenous and native groups, anthropologists and scientific communities, who argued that 'ethnic groups are categories of human invention, not given by nature'.[8] Scholars challenged the 'rhetoric of preservation, time pressure and alarm' and some Indigenous peoples called it the 'Vampire Project'.[9] The 'sampling' of communities selected for both their rarity and use-value seemed to many Indigenous and native communities to be a type of biocolonialism.[10] Critical voices pointed out that 'presently structured, many interactions of genome scientists with Indigenous peoples see them as a means to an end'.[11] Communities also opposed seeming 'constructions of who could speak for the concerns of Indigenous people'.[12] Many communities felt that profound coproduction was missing throughout because the project still insisted upon reading identities and categories from the data.

The project continues despite not being funded or supported in the same way as the HGP, and now consists of 51 populations from around the world.[13] In 2005 Cavalli-Sforza robustly defended the project in an update for *Nature* and outlined the ways that such work helped biomedical research and work on evolution.[14] Cavalli-Sforza suggests that claims about commercial exploitation were wrong by saying 'DNA samples would be provided only to non-profit-making laboratories' (p. 333). This does not really answer the problems about collection, attitude, rhetoric, the manipulation of resource and notions of biocolonialism. He also argues that accusations of 'scientific racism' were wide of the mark because 'half a century of research into human variation has supported the opposite point of view – that there is no scientific basis for racism' (p. 333). This misses the point that 'scientific racism' is not the same as scientifically-based racism, racist science, or the wider reporting of science in the media.[15] Problematic attitudes including the sense that science is somehow apolitical and therefore is impervious to challenge (what Whitt terms 'the neopositivist assumption of value neutrality') contribute to structural and specific problems in practice.[16]

The HGDP highlights an ongoing division between the genetic scientific community and Indigenous and native groups, particularly with relation to the value of genetic work and the relationship between DNA and the human past.[17] This is often to do with unconcious conservative attitude and a misreading of historical difference. Amade M'Charek points out that the HGDP reifies one version of history and origin. It suggests a snapshot of isolated tribes that may not have been so forever (p. 13). Some groups have become isolated only through colonialisation or mass migration. Therefore, the project introduces a periodisation (then/ now, modern/ ancient) to genetic data which is false. It suggests a type of purity is to be sought in knowing the historical human, and further, it seemingly imposes ideological models upon Indigenous data and assumes authority over it. The distinction made between 'other' and the hybrid genetics of the centre contributes to a continued marginalisation of these native groups and one which reifies this as historically significant and insuperable. In terms of decolonisation, the HGDP seems to participate in continuing repressive frameworks whilst appearing to offer aid and support. M'Charek reminds us that 'Race does not inhere in skin color, physical characteristics, a palmar crease, DNA, clothing, national identity, or the like. It is a configuration, an effect of relations between differences. It is thus that race is a relational object'.[18] The problems of discussing race within genetic science often arise from divorcing this social understanding of the category of race from discussion.

The Havasupai Tribe and Aznick-1

Many Indigenous and native communities have been suspicious of DNA data collection for several decades, for a range of reasons. Primarily there are concerns regarding the historical collection of genetic information and the ways that it has been used. In one of the most important cases, the Havasupai tribe sued Arizona

State University for using genetic samples in a range of studies rather than the initial one they were collected for.[19] The original research had concerned the high incidence of diabetes amongst members of the tribe; researchers later used the data collected to study and publish on a range of controversial and ethically problematic areas. The settlement in 2010 recognised that the principle of informed consent had been violated.[20] The misuse of Havasupai DNA echoed multiple cases around the world in terms of treating these often disadvantaged and marginalised communities poorly, not including them in partnership, and representing them with little or no knowledge of their history, collective identity, and traditions.[21] The Havasupai case demonstrates the damage that can be done by researchers not recognising the significance of the data they are collecting (or ignoring it), and how fragile trust is easily damaged.[22]

Unlike Richard III (see Chapter 1), the remains of the boy known as 'Anzick-1' were reburied in 2014 relatively quietly. The child's remains were discovered on a farm in Montana in 1968, buried with many artefacts identified as part of the Clovis culture. Named after the family who owned the farm, the remains, which had been buried between 12707 and 12556 years ago, were sequenced for DNA in 2014 before being reinterred. The investigation by the pioneer of Ancient DNA work, Eske Willerslev from the University of Copenhagen, suggested that Anzick-1 was both a direct ancestor (through Y-Chromosone DNA lineage) of contemporary Indigenous communities in North America and also had Asian ancestors.[23] The work decisively challenges that the so-called 'Solutrean' hypothesis that Native Americans moved to the continent from Europe.[24] The study had this key assertion at its conclusion:

> The Anzick-1 data thus serves to unify the genetic and archaeological records of early North America, it is consistent with a human occupation of the Americas a few thousand years before Clovis and demonstrates that contemporary Native Americans are descendants of the first people to settle successfully in the Americas.[25]

This combines the cutting-edge science with historical interpretation, making a clear assertion about events in the past that has contemporary political resonance. The sequencing of the DNA was part of the final process of investigating the remains before they were reinterred. The ceremony was brief and low-key, but provoked debate about the use and excavation of human remains. Shane Doyle, an academic who ensured the reburial would happen, argued that "'From a tribal point of view, this is a big part of reclaiming our history, reclaiming our dignity for our kids.'"[26] The same article in the *Billings Gazette* quotes Armand Minthorn of the Umatilla Tribe (now located in Oregon): "'These are our ancestors' remains, they are not artefacts. I hope that the people who come after us remember this, as well.'" The reinterring of Anzick-1 is part of the on-going worldwide Indigenous reclamation of the physical materials of ancient memory. There are many points of conflict, including Kennewick Man in the USA and Mungo Lady in Australia, where DNA research has led to some resolution and enabled tribes and native communities some restitution and partnership.[27] The language used by researchers has changed a little due to their engagement with these groups, and there is some recognition of their concerns, although there is much work still to be done.[28]

Indigenous and native scholars have continued to challenge the assumptions of current genomics, in particular the assumptions around using samples. These scholars have criticised the 'bone-rush' in which 'irrevocable decisions continue to be made about the sampling of ancient specimens, guided by the immediate research interests of a few'.[29] Some researchers have sought to outline their own processes and ethical procedures around the collection and manipulation of Indigenous genetic material, although they recognise that 'the participation of Indigenous communities in genomic research will require the *recognition* of their rights and interests in genomic data and consequently will require *radical* improvements in trust, equity and accountability'.[30] Multiple instances of contemporary biocolonialism around the world underline the assertions of native scholars that genomic science is allowing the Indigenous body to be once again annexed.[31] Jenny Reardon and Kim TallBear argue that 'Native American DNA has emerged as a new natural resource that Native peoples possess but that the modern subject – the self-identified European – has the desire and ability to develop into knowledge that is of value and use to all humans'.[32] They demonstrate the ongoing assumptions made by academic institutions and practitioners, and the continuing problems caused by working within particular frameworks that continue colonial ideologies. The case clearly suggests the need for radical reform and critique. There is a need for further understanding of what Reardon and TallBear understand as 'the relations between whiteness, property, and technoscience' (p. S235), and particularly the ways in which Indigenous and native peoples are figured as 'ancestors' and therefore scientifically – and historically – othered. In response, native and Indigenous groups have challenged the purpose and consequence of data collection, arguing that research design does not include them and also actively works to exclude them.[33] Many Indigenous groups are seeking 'genomic data sovereignty' as part of their self-governance and ownership of their past, but researchers circumventing or ignoring tribal structures can still relatively easily undermine this.[34] There is a strong critique in various communities of the general purpose of genetic science and the ways it ignores Indigenous knowledge.[35] Indigenous populations are marginalised through not being involved in the direction of scientific research, or, as Kathryn Milligan-Myhre describes her time as an undergraduate, 'I was not studying questions that were important to Inupiaq'.[36] The best practice so far developed is to seek partnership when developing research outcomes, ensuring that research is not imposed upon the community.[37] However this stipulation is often evaded or ignored, and as a recent study argues

> Numerous research projects, genomic or otherwise, exhibit enduring negative effects on Indigenous Peoples, minority populations or socially disadvantaged groups owing to under-representation, lack of informed consent, lack of consultation, misinterpretation and/ or misuse of sample and data.[38]

At present, too, legislation around the world only really relates to communities that are organised and recognised as sovereign, so marginal groups in Brazil or Colombia, for instance, have little protection. As Klaw et al outline, Indigenous peoples around the world have developed their own policies relating to genetic work including ethical guidelines and advice on best practice.[39] Important initiatives have been set up to enable communities to lead on and partner with genetic research, including work undertaken by the Native BioData Consortium (USA), Alaskan Area Specimen Bank (USA), H3 Africa Group, Te Mata Hautū Taketake-Māori & Indigenous Governance Centre (New Zealand), and the Australian National Centre for Indigenous Genomics.[40] Some areas, such as Newfoundland in Canada, have dedicated laws and state mandated ethics boards devoted to working in partnership with Indigenous communities to help protect their data from being misused, although this type of legislation is relatively rare and there are criticisms of the policy outlines given their generalised and centralised nature.[41] DNA investigation allows a more complex picture of population history and development.[42] However the lack of diversity within major genetic databases means that the information is not thorough and leads to continued marginalisation of non-European communities.[43] Increasingly, although slowly, this work is investigating a diversity of range, investigating populations from Cuba, Latin America, and Asia; however 'discovery efforts in non-European populations remain limited'.[44]

There are further criticisms of commercial, consumer-focused DNA collection and use relating to Indigenous identity. Indigenous and native communities have consistently rejected the idea that identity might be simplistically conferred through genetics (see below).[45] This means that Direct-To-Consumer services suggesting that they can 'confirm' genetic identity are misleading as they work on a 'reductionist interpretation of tribal identity' and ignore the 'complex social, political and historical narratives that comprise such identities'.[46] These companies simplify 'Native identity to a collection of biological determinants, superseding important cultural and historical considerations'.[47] Scientific criticism of the DTC representation of ethnicity underlines that race is not strictly a biological category:

> Current understandings of race and ethnicity reflect more than genetic relatedness, though, having been defined in particular sociohistorical contexts (i.e., European and American colonialism). In addition, social relationships and life experiences have been as important as biological ancestry in shaping individual identity and group membership.[48]

Many participants have used these services to claim heritage, leading to 'indigenisation' and wrongful assertions of identity.[49] Membership of Indigenous groups in some areas of the world confers some social and legal status (although others are aggressively marginalised, as is the case in Brazil for instance). That notwithstanding, communities have worked for decades to defend the principle that genetic identity is not the single-defining element of membership, and indeed for many groups is not deemed relevant. Indigenous peoples around the world have multiple and complex understanding of what it means to be part of, and accepted within, the various communities. As Kim TallBear has argued, identity 'gets represented as this purely racial category by some of the companies marketing these tests. The story is so much more complicated than that'.[50] DNA evidence is marketed by the DTC companies as defining and enabling an identity, and the impact is such that so-called 'Native American' and Indigenous ethnic characteristics are identified and organised by the companies, rather than the communities. TallBear outlines that genetic information is utilised by tribes in a 'very particular social and historical context, one that entangles genetic information in a web of known family relations, reservation histories, and tribal and federal-government regulations'.[51] She argues further, Indigenous identity is 'not just a matter of what you claim, but [...] of who claims you'.[52] This complexity is ignored in the rush to confer ethnic identity and heritage undertaken by the Direct-To-Consumer companies.

Elizabeth Warren and the Political Use of Genetic Identity

Concerns about the conflation of seemingly definitional genetic 'ethnicity' with political identity gained media prominence in late 2018 due to the actions of Elizabeth Warren, United States Senator for Massachusetts. For many years political opponents have attacked Warren over her claim to Cherokee ancestry. It was alleged that she used her ethnic background and heritage to gain preferential treatment in her legal career, which she denied. After extensive personal attacks from President Donald Trump about this issue, Warren decided to use a DNA test to 'prove' her identity ahead of the beginning of her campaign to become the Democratic Presidential nominee.[53] Indeed, Trump had goaded her to have such a test to prove herself. In October 2018 Warren had her DNA analysed by an expert at Stanford University, Carlos D. Bustamante, who reported that Warren's genetic profile 'strongly' suggested that she had an 'unadmixed Native American' ancestor 'in the range of 6-10 generations ago'.[54] This was subsequently widely reported as meaning she was 'between 1/64th and 1/1,024th Native American'.[55] Warren's genetic identity was worldwide news given her status, and the results were reported as a challenge to Trump's belittling of her. Warren did distinguish clearly between her family history and any claim to tribal identity, but this subtlety was broadly lost in the press coverage, which instead stressed the fact of her 'Native American' identity being proved by the results. The case was very high profile, and, clearly, this was intentional. Warren was seeking a political advantage in order to nullify President Trump's insults and turn them against him by claiming she had proof that he was racist.

Indigenous groups quickly pointed out that such genetic inheritance is not the only part of identity, and furthermore that the science behind these indications could be imprecise. The Cherokee nation released a response firmly stating that 'A DNA test is useless to determine tribal citizenship [...] Using a DNA test to lay claim to any connection to the Cherokee Nation or any tribal nation, even vaguely, is inappropriate and wrong'.[56] Chuck Hoskin, secretary of state for the Cherokee nation, was careful to underline the fact that genetic science currently played no part in Cherokee identity: 'DNA is completely irrelevant to the process [...] We have a very rigorous process, and it's about documentation and tracing back'.[57] Science commentators reiterated the limits of such work and criticised how it was reported around the world.[58] Indigenous scholars and experts moreover argued that Warren's publicity stunt was against tribal interests, communicated the wrong ideas about tribal identity, and continued the implementation of science on behalf of the mainstream. Seeking to 'define' historical identity through genetics has been interpreted by tribal peoples as white privilege since it rests on the ability to confirm 'settler-colonial definitions of who is Indigenous'.[59]

Warren's case demonstrates again the public importance and influence of commercial DNA tests. Discussion of ancestry and genetic background has become political currency of a kind. Whilst Warren did not take a commercial test she anticipated that the public would comprehend the work that was done in identifying her genetic background. Part of the point here is that DNA replaces other types of historical investigation, and allows Warren to claim a truth about herself, despite the attendant controversy. It is a different way of knowing herself, and her relation to events and people and communities. She becomes genetically historified, given extra dimension, that of a particular type of connection to communities and peoples in the past. The purpose of the DNA reveal seems to have been to 'locate' Warren precisely, and to gain political advantage from that. However, the move did not work politically, and indeed backfired; Warren apologised to the Cherokee nation and continued to apologise throughout the opening of her presidential nominee campaign.[60] In this moment, then, genetic identification of history and heritage was resisted. Yet the arguments around Warren's genetic identity and claiming of community are a version of those that have been developed for many years regarding scientific definitions of race more widely.[61]

The Return of 'Racist' Science

Much disquiet about the imbrication of genetic science and race is due to a general post-war concern with eugenic science coupled with clear examples of problematic practice. The link between genetics and race has a long and often unpleasant history, from the work of Francis Galton in founding eugenic science to the later views of James Watson on race and intelligence (see discussion in chapter 1).[62] However the past two decades have seen a rise in racialised interpretations of science. There has been a clear increase in hate crimes and racialised discourses that have been motivated by the application of scientific information for ideological purposes.[63] More often than not the science used is genetic science, given that this type of work is popularly agreed to outline attributes relating to race and ethnicity. As Adam Rutherford argues:

Reluctance by scientists to express views concerning the politics that might emerge from human genetics is a position perhaps worth reconsidering, as people who misuse science for ideological ends have no such compunction and embrace modern technology to spread their messages far and wide.[64]

Genetics has been appropriated by far-right groups in their arguments about legitimacy, purity, authenticity and nationalism.[65] Right-wing groups in the USA are using commercial tests as part of a desire to prove racial purity.[66] These groups, and those figures who use science to support theories of legitimacy and purity, are utilising the cultural power of DNA – its scientific mandate to truth, in popular usage – to develop new arguments about race and identity. They are seeking a means to reach beyond the modern world and connect to 'racially pure' lines through history.

Genetic difference remains an important popular way that people define race throughout societies.[67] The significant implications for cultural and social understanding of race due to developments in genetics have been clear for some time.[68] Genetics at present seems to explain ethnicity and secure racial identity, and most commercial tests will outline these categories with much confidence. In his book *Tribes*, which seeks to understand contemporary political division, British politician David Lammy describes taking a DNA test and the way that it connected him to a history he had never had before.[69] This story of using the Direct-To-Consumer tests to connect to a multiple, diverse, and complicated personal history is common. The complexity of the situation regarding genetics, in terms of opening up the story (in Lammy's case) or closing it down – in the case of other users – demonstrates the various ways that DNA tests might be interpreted and the multiple ways the results can communicate or shape present identity in relation to individual (and group) history.

Historians and archaeologists who use this material need to critique, challenge, and debate the assumptions and ideologies inherent in the data. Particularly, they need to consider the underlying assumptions behind the work that produced the results. This might be the valuable interface between humanities and scientific approaches to the past via DNA. For instance, archaeologists have criticised the use of genetics in archaeological practice. Articles have suggested that aDNA research is too rigid in its understanding of human interaction, mistakenly imposing a template upon diversity and complexity.[70] Others have been more strident, calling out what they see as 'biodeterminism', particularly in the work of David Reich. Michael Blakey describes what he calls a 'near fetish of DNA' in 'Western science and society' and attacking the 'excessive authority ascribed to genetic explanation'.[71] Blakey warns archaeologists about working with scientists whose approaches are predicated upon problematic and unspoken deterministic ideologies. This need for a feedback loop, for historians to provide a critique of longstanding institutional and social biases in the production and interpretation of data would be one step towards decolonisation of DNA. What is needed is the creation of an historical nexus of interpretation for how and why genetic information is generated and understood; this should be a process of collaboration and critique, providing a necessary challenge to epistemologically-based assumptions.

Genetic Nationalism

One of the concerns about DNA usage in relation to history is how it intersects with 'genomic liberalism' and fantasies of multicultural inclusivity. Specifically, this pertains to a sense of aspirational futurity predicated upon an uncritical sense of genetics as a positive science. Jenny Reardon has outlined 'genomic liberalism' as a performative attempt to present genetic science as inherently anti-racist.[72] In her critique Reardon challenges this discourse, particularly in relation to how such approaches 'may play key roles in the individualization of race' (p. 41), prioritising self-identification rather than wider social understanding of anti-racist action. This type of discourse can be seen in the promotional work of Ancestry.com, for instance, in its attempts to promote DNA ethnicity as a means for understanding 'beyond 'race. The promotional film 'Declaration Descendants', for instance, re-enacted the famous image of Thomas Jefferson presenting the Declaration of Independence to Congress in 1776.[73] The re-enactment brings together descendants of the original figures in the image, the founding fathers of the country. The historical dissonance of the film is that the contemporary image includes women, people of colour, and descendants of Sally Hemmings, Jefferson's slave (a relationship only proven and accepted due to DNA evidence, see Chapter 1). Ancestry's aspiration was to articulate a foundational diversity enhanced and enabled through genetic knowledge. Vineet Mehra, Chief Marketing Officer, argued that

> Diversity isn't just something we value as Americans; it's quite literally part of who we are [...] one of the most powerful things we can do is to show how connected we really are [...] we share an inherent need to know who we are and how we're related – it can change how we view the world and how we view our future.[74]

Mehra connects genomic liberalism with aspirational marketing, combining revision of the past with a commodified aspirational genetic diversity. Ancestry presents an aspirational way of thinking about ethnicity and family history, and this inclusivity is predicated upon a way of conceptualising a future space where 'race' is replaced with 'diversity'.

In contrast, much discussion of nationalism and DNA is reductive and seeks to isolate precise and cliched characteristics. In July 2016 *The Telegraph* (UK) ran an article entitled 'How British Are You?'. The article reported an Ancestry.com study that looked at 15000 users of its DNA sequencing service. Amongst other things the study looked for DNA matches with 'Anglo-Saxon', and this is the definition of 'British' that the *Telegraph* decided to emphasise. The average resident of the UK (that takes this test) is 36.94% Anglo-Saxon, 21.59% Celtic, 19.91%

Western European, 9.2% Scandinavia, with other elements from the Iberian Peninsula, Italy, Greece, with smatterings of European Jewish, Finland, Russia, Eastern Europe. The article demonstrates the way that DNA sampling for family history purposes is being turned into data. Participants might discover their individual ancestry, but their information is being read back into the archive to make suggestions about current population definitions. Without bioarchaeology – work on population genetics, on sampling of ancient DNA, on language formation – the data would have no way of being understood. It is being 'read' in relation to ethnicity and made to mean. This reading of the contemporary is dependent on an understanding of DNA in history, a conceptualising of ethnicity at a point in the past that is dependent on historical interpretation and definition of that ethnicity.

The article suggested that people from Yorkshire were the most 'British', and was illustrated with images of cricket, tea, and various Yorkshire sights such as the Ribblehead Viaduct. It is a strange article that seeks to define a type of Britishness even when the data appears to be pointing toward a much more complex sense of identity. Quotes from various Ancestry figures back this up as they point out that 'The UK has been a cultural and ethnic melting pot for not just generations, but centuries'.[75] Coming a month after the controversial Brexit vote in the UK this report was covered in the News section of the paper and testified to a continuing interest in nationality, ethnicity, and DNA. It also shows a tendency to associate a type of 'Britishness' with Anglo-Saxon genetic identity. The Anglo-Saxons are an immigrant group, of course, but also one associated with establishing the nation of 'England' in terms of land organisation, institutional development, and government. Whilst we might expect that the press interpretation of the data might lean to the popular, Ancestry's own reporting participated in lazy clichés: 'Live in the East of England and always wondered why you have a strong penchant for pizza, pastries, and gyros? It might have something to do with the East of England having the most Italian/Greek (Southern European) ancestry (2.53%) and Western European (French/German) (22.52%) ancestry, as well as the highest amount from the Iberian Peninsula (Spain/Portugal) (3.43%)'.[76]

The article and the Ancestry blog read contemporary identity, via DNA sequencing and markers, in relation to historical ethnic groups. In particular, the Anglo-Saxon group (*c*.450-1100), and Celts, are articulated as somehow connected to communities in the now. This is an amazingly complicated thing to assume, defining a body – a set of cells, a collection of genes, some genetic information coded into a sample – physiologically, geographically, nationally and historically.[77] The complex version of nation identity that is expressed here (particularly, an *English* nationalism) is dependent on contemporary clichés and identity codes (tea, pizza) being reinforced by genetic science and popular historical understanding. Somehow, encoded within our contemporary bodies, and identifiable through technological sequences, a type of national identity is woven into the very building blocks of our cells. DNA information is something that makes the contemporary human multitemporal, complex. Our bodies cannot lie to us, and the evidence to be found within them presents us with irrefutable proof of *something*. However, what that is will be immediately interpreted through a contemporary lens, can only be understood as affirmation of a set of identity signifiers that are clearly nothing to do with anything physiological.

Cheddar Man and Nationalist History

The conflation of ethnic 'Britishness' with genetic understanding of the past can be seen in the case of 'Cheddar Man'. This case shows particularly how genetic work can seek to be politically revisionist. Remains of a hominid were found in caves in Cheddar Gorge, UK, in 1903-4.[78] The bones of what became known as 'Cheddar Man' date from the Mesolithic period (around 7000 BCE) when the area known as 'Doggerland' connected Great Britain to Europe. At present kept in the Evolution gallery at the National History Museum, they are some of the oldest known hominid remains in Europe. A reconstruction of the head of 'Cheddar Man' had been created by the University of Manchester, presenting the figure as pale skinned with dark hair. In 2017 further genetic work was undertaken on the remains.[79] The results suggested that the hominid had dark skin pigmentation, dark hair, and blue eyes. The notion that 'first' populations of the UK were dark skinned was incredibly controversial, and the data had an impact around the world.[80] That ancient Europeans had dark skin pigmentation had been relatively well known within bioarchaeological circles for some time.[81] The genetic work on 'Cheddar Man', though, was a breakthrough in terms of wider public knowledge. In part this was because of heightened media interest. The entire process of the genetic analysis was followed by a documentary team and the resulting film, *First Brit: Secrets of the 10,000 Year Old Man* (Channel 4) was screened in 2018. The case of 'Cheddar Man' sits at the nexus of representation, materiality, public understanding of archaeology, the ethics of displaying human remains, and the performative nature of science (Figure 3.1).

'First Brit'

First Brit is many things: a documentary showing the practice of a-DNA investigation and archeogenetic work; a piece of public science education; and a political contribution to debates about ethnicity, race, and the popular understanding of the human through genetics. The film begins by claiming 'science is about to reveal where we come from. And who we really are', and how 'it may be that we have to rethink some of our notions of what it is to be British'. This revelatory sense that the data may shift national identity is important to the way that the documentary presents historical evidence, 'shedding light on unanswered questions'. The audience is reminded that 'It is a story that begins over 10000 years ago, before Britain became an island and our first ancestors arrived'. This notion of a pre-'British' past, a prehistory that is before nation, is part of the revelatory aspect of the documentary. This revelation is somewhat anxiously articulated at times, as participants remind the audience that 'we're not conjuring this out of nowhere, we

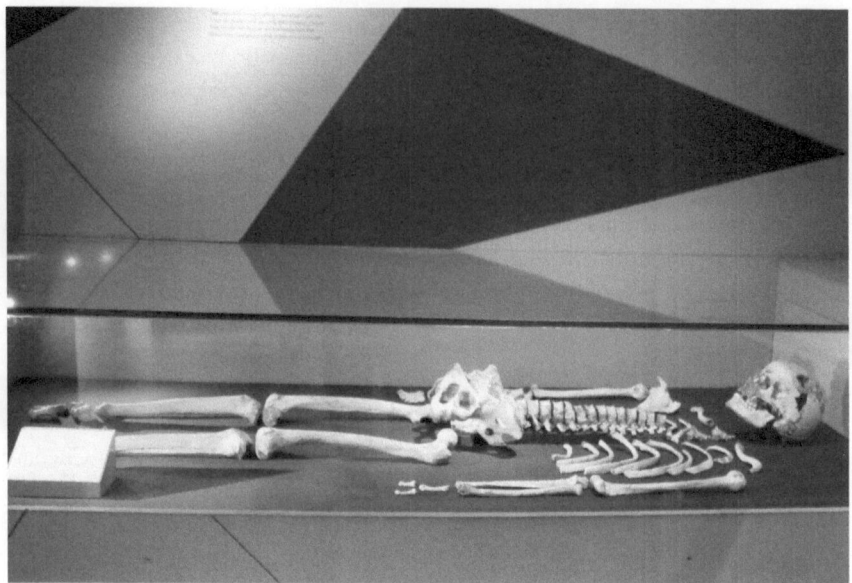

FIGURE 3.1 Remains of 'Cheddar Man' in the Natural History Museum, London

Creative commons Attribution-Share Alike 4.0 International Licence, https://commons.wikimedia.org/wiki/File:Em_-_Cheddar_Man_-_1.jpg.

Credit: Emőke Dénes.

really do have scientific data'. Consequently, it is asserted, this information will effect an epistemological change, 'allowing us to come face to face with the first Brit'. The documentary, then, is presenting a scientifically-derived revisionist argument about historical event and the heritage of nation. In doing so it is challenging ways of thinking about Britishness, particularly in relation to race and to Europe.[82] It explicitly argues for an encounter with 'First Brit', genetic science enabling the contraction of historical time to enable a face-to-face meeting.

At the opening of the documentary the remains are laid out and two scientists examine the skull, talking about accessibility. The skeletal remains – referred to often as 'it' by the voiceover – is in fact a 'specimen', in the repeated term used by one contributor. There is no discussion of the ethics of the work that is being undertaken, of displaying and working on humans.[83] Whilst this project no doubt was subject to ethical and legal strictures, they do not appear onscreen, and the historiographical process of the television programme is compromised. What is the ethical duty of a documentary in presenting human remains and considering them in heritage and scientific contexts? How should television present the investigation into the past through such material analysis? These questions seem particularly apposite through the sequence in the programme in which the sample is taken. The skull is drilled with a very small hole and the process is described in precise detail: 'So that out from that hole will pour this very white bone powder [...] from that we will be able to pull out the DNA'. This intrusive action – the drilling – enables a further invasion of the body, as the stuff of life – the genetic code – is revealed. The audience is reminded of the revelatory quality of the investigation – 'These few milligrams of bone powder could contain secrets hidden for 10000 years' – whilst the quietly whining drill works on the skull. The action is both intrusive and clean, the bone powder softly falling away. This reification of the genetic scientist's power in revealing 'secrets' through their interrogation of the physical continues throughout the programme. Similarly, the 'sample' now becomes abstracted from the skeleton – the skull – and increasingly loses any agency as the dead, or as a former human.[84] The scientist intervenes in the physical integrity of the human remains, turning it from something solid into a 'sample' of fragments.

The DNA work itself is undertaken behind closed doors. This is literally the case, and is emphasised, as several shots show the key geneticist moving from the museum space through doors into a lab environment, away from public display. The genetic investigation is undertaken outside of the heritage environment, outside of 'public' scrutiny. The editing at this point demonstrates the movement from seen to unseen types of knowledge. The heritage space of the museum is exchanged for the clinical atmosphere of the lab – at times even filmed from afar (the above image is shot through glass) to emphasise the 'difference' of this type of activity. The shots here emphasise a kind of precision and cleanliness – there are shots of plastic gloves and clean suits being put on, discussion of contamination – and technical innovation (lingering on machines, computers, sterile spaces). This section presents the lab as an antiseptic, non-human space in which the sample – the skull – stands out as something organic, old, and strange. The sections shot in various labs emphasise analysis as something clinical and precise. The 'historical' work being done here is technical and clean, part of a wider practice of investigation that is different from that undertaken by historians and mainstream museum curators. Each of the interviewees emphasises the *newness* of what is being done – claiming a modernity for this investigation of the ancient past. At its most extreme this includes reconstruction work, which depends on the skull being scanned by a 'hi-tech scanner' designed 'for use on the International Space Station'. The project therefore submits the aged organic matter to a measuring instrument so modern it is not designed to be used on planet earth.

The specific genetic work filmed inside the DNA lab is important to consider in some depth, since it is one of the first times that the *process* of ancient DNA analysis has been presented onscreen. Similarly, the emphasis on taking the investigation into the 'scientific' space away from the rest of the museum seems to suggest that certain types of knowledge might be revealed there. The practices are not explained in any detail, just a set of images of 'scientific' moves by figures in lab coats that signify 'bio' investigation. The VO assures the audience that the genetic work is done in a 'hi-tech DNA sequencing lab' filled with 'state of the art £200000 machines [that] can crunch tens of millions of pieces of data, a process that used to take several weeks and still takes many hours'. The machines' value confers a kind of gravitas, but just as important is the scale of the information collection and the new speed of work. The human is a hugely complicated species, but the technology is able to deal with this by 'crunching' the information. Indeed, the actual innovations of Ancient-DNA analysis are never explained, and the process is not outlined – the audience is simply told that 'the sequencing' will reveal much, as '10000 year secrets are about to be spilled'. The VO talks of how Cheddar Man may 'give up his secrets' to the lab, but little is discussed of how this might work or why DNA is so revealing. The scientists simply take the sample into the lab, do their work, and the 'secrets' emerge.

The programme focuses particularly on the pigmentation issues and the inter-relation between genes and skin colour. There is some description of how comparative DNA work might inform researchers about eye colour and skin colour, although the work is highly simplified. Further shots show comparative data analysis, modelling, and population statistics. The scientists work collaboratively, suggesting links between their individual and others around the world that have been sequenced. This comparative work is well described and outlines the way that understanding the DNA of individuals is something that considers wider populations. That said, there are many shots of people looking at computer screens and working on laptops. Knowledge is suggested to be obvious and clear but is not shown particularly or shared with a viewer – only conclusions and assertions (although in careful language). Often screens are presented – tables of data and graphs. This historical investigation is based on 'reading' data and analysing figures. Again, this may be the first time such investigation has been presented on television and particularly it is important for suggesting that 'historical' investigation of this type can force genetic material to 'give up' secrets. The information is presented as data, generated by various collaborative investigative practices. The non-scientist in the group, an archaeologist, provides only one insight in the programme which is based on his reading of the DNA sequence, so even 'traditional' roles are becoming part of the scientific investigative model.

Part of the process of recovery and reordering the remains was the creation of a new model of 'Cheddar Man', created from a combination of advanced scanning techniques and the interpretation of genetic data. The reconstruction of the face and shoulders is an artistic practice, and the physical work in the studio is highlighted. The portrait is three-dimensional, an artist's impression worked up from the information gleaned by the geneticists. Whilst much of the facial reconstruction depends on work on the skeleton, the pigmentation, hair colour and eye colour relate directly to the genetic investigation. The Kennis brothers, who made the reconstruction work hard on the pigmentation and build the 'face' in various ways. The scientists talk about how exciting and unusual it is to get something 'physical' to look at; the artists are excited to 'reveal' the 'distinctive' face. The work that the two artists do leads the geneticists to reflect upon the link between 'data' and materiality: 'To have done the genetics on someone [...] and then to actually see that made into flesh is amazing'. Their piece of work 'makes' 'Cheddar Man'. This link between the data and the 'real', or at least the constructed real, suggests that the reconstruction is more valid than the original or the data: 'he's alive, he's a person now, he's not just bones'. The visual is important in making 'him' somehow real, and to communicating the message that the DNA work discovered. The documentary uses the reality TV structure of the reveal and the make-over, as 'Cheddar Man' *becomes* through the processes we see. This generic template allows this shift from 'sample' to 'individual' to 'a person', as the figure is somehow dragged into through history into the present. He is literally 'revealed' at the end of the show, as the artists stage an unveiling in the hall of the Natural History Museum. Motifs of art, performance, science, reality documentary and staging come together here, a revisionist account of human development articulated through the revealing of a sculpture.

Science, Nationalism, and Heritage

The final sequence of the programme reflects upon the relation between then and now, suggesting a key link through history via genetics. The work that had been done relating modern 'Brits' to Cheddar Man's DNA reveals 'something surprising', a DNA echo that means '300 generations on, we're all a little bit Cheddar Man'. The use of 'we' and 'our' is intentional throughout the script, emphasising a link between then and now, and engaging the audience too with this connection. The VO concludes forcefully: 'DNA science has truly unlocked the 10000-year-old secrets of his identity and given a face to our ancestor, the First Brit. He marks the beginning of our national story and has left a genetic legacy that we all still carry with us today'. This shift in historiographical and historical awareness is made possible by genetics. The reference to a 'national story' suggests a sense that the shift from prehistory into nationhood is being made with the 'recovery' of Cheddar Man. He is presented as the 'First Brit', an ethnically and historically meaningless appellation but one that emphasises a link to contemporary populations. The point of this is to reinforce the 'we are all a little bit Cheddar Man' – and to underline human genetic diversity, the pointlessness of contemporary identity signifiers. 'Our' nation (they assert) began with this original figure – only now understood through the intervention of genetics – and we all owe him something intangible. We are all 'part' of him. The impact of the scientific work is historiographic and revisionist: 'it shows up that these imaginary racial categories that we have are really very modern constructions, or very recent constructions, that are not really very applicable to the past at all'. DNA work on material from the past can be corrective to the present. It can work to undermine and deconstruct the contemporary sense of identity and selfhood. It also provides a disconnect between then and now, insofar as it suggests that contemporary 'constructions' actually breaks the contact from the 'real' past.

The programme is making a conscious political intervention. This is the reason for the focus on skin pigmentation and the continuous emphasising of a link between then and now. The work on pigmentation and skin colour, and indeed on physiology and physical characteristics, is part of a much wider articulation of results intended to prove a series of hypotheses regarding migration and the development of human populations in the area that became Britain. The naming of 'First Brit' consciously echoes and inverts the name of the racist rightwing organisation 'Britain First'. Made in the immediate aftermath of the Brexit vote for the UK to leave the European Union (2016), the emphasis on historical difference and a challenge to notions of innate, natural 'Englishness' are clearly politicised.[85] One of the contributors from UCL, Yoan Diekmann, adds that the type of work that is being done is fundamentally revisionist due to its scale and breadth: 'the historical perspective you get is that things change, things are in flux, and what may seem as a cemented truth [...] through time is not at all something that is an unmutable [*sic*] truth'. Diekmann suggests that taking such a long view of human data shifts the way we might think about 'truth' and identity; the sense of mutability is given to the researcher through 'historical perspective'. The programme also emphasises the 'new' techniques of investigation that are being used to reveal information about prehistory and humanness.

The revelations of the scientific study and the documentary were discussed widely in the press. There were articles across the tabloid press in the United Kingdom, reflecting huge public interest in the revelations. *The Sun* illustrated the 'unprecedented examination of his DNA' in this 'pioneering research' and emphasised how 'the findings are a dramatic departure from earlier reconstructions'.[86] Reporting in the *Daily Mail*, Colin Fernandez similarly emphasised the 'Only now with cutting-edge DNA and facial reconstruction techniques can we see for the first time the face of this 10,000 year old man, and ask how 300 generations later he relates to us today'.[87] It is illustrative to look at the huge number of comments (1400) on the *Daily Mail* article, which itself has been shared 44000 times, to gain a sense of the impact of this work. Much of the commentary is about overreaching the facts, fake information, misinterpretation, or simply shows misunderstanding of the fundamentals. In some ways this is a case-study for the problems about reporting challenging scientific research in the popular media, and indeed the importance here is that DNA investigation into the past is still complex and difficult for the lay person to gather, or for journalists to narrativise (although the tabloid articles are admirably lengthy and have much information in them). The comments show the diversity of opinion relating to the information about Cheddar Man, but also demonstrate the ways in which the new interpretation is clearly not neutral. Comments veer from 'What a load of baloney' to 'Fake'.[88] Throughout the comments is a strand of aggressive disbelief arguing that the genetic science is being used for a particular agenda:

> 'One individual. One. And 76% chance … and with support from Ch4 the most biased news channel that is out to demonise all white people and their culture …'

> 'Aha, so again so-called 'scientists' see political correctness as more important than rigorous scientific research. Maybe they can see the millions in Government funding for 'Man-made global warming' research start to dry up'

'Left wing claptrap'

'This MUST be 100% TRUE and ABSOLUTELY NOT agenda driven'

'Ha, ha, No surprise that the right-on scientists said it's black. Any colour but white will do. All part of the new order narrative'

'These 'reconstructions' tell you more about contemporary 'acceptable' attitudes than they do about the past – and always have done'[89]

One user claimed about the story, 'No way is this true. It's PC gone mad'.[90] To an extent this user is right, inasmuch as this work isn't 'true' (and seeks to avoid a binary of true/false). More profoundly this user's insight demonstrates how historically embedded the genetic investigative moment is. Cheddar Man has been reexamined in the context of political, institutional, and cultural structures. The questions that have been asked of these pieces of bone are particular to this time and place, with genetic investigation simply part of a suite of investigative technologies that carry historiographic function as well as a historically situated ideology. As such, it is 'PC gone mad', insofar as it is precisely located at this moment in order to make assertions germane to the contemporary understanding of the past and its relation to the present. The science of this work, as is the case with all the work that is being done currently on ancient DNA, is historically contingent and part of a wider infra-structure of investigation into the past. The infrastructure of investigation includes universities, museums, laboratories and popular historical institutions. Like Richard III (see Chapter 1), 'Cheddar Man' is a nexus for investigative modes of pastness, each of which reveal much about contemporary ideas of the past and the ethics of heritage. 'Cheddar Man' is made to mean for us now and the process of his trans-lation into the modern world is the genetic work that is done on his dormant DNA. This individual-species ('Man') had been in a kind of drag and performing for decades in the museum. The desire to know him and make him again has changed his dimensions, his manifestation, his meaning, and his pigmentation. The memorial practices that are deployed lead to him being re-known, rendered again in multiple ways that are little to do with his original life. He is reborn, remade, rethought, re-understood. He is prehistoric, that is, pre-history; beyond our understanding and psychology, surely, and beyond our contemporary definitions of race, class, gender, biology, outwith our language, social organisation, definitions of self, models of rationality, before nationhood, previous, proleptic, challenging, progenitor, first, literally an alpha male. All the while DNA dwelt in his drying and decaying body, significant despite their death, detectable eventually. The 'secrets' were there to be given up when addressed in the right way, when asked the correct questions. They survived his death and existed as potential for them to be reanimated and reinscribed.

The signifying bones we know as Cheddar Man move from being 'real' to being dead and absent, then they are found and imbued with a different, educa-tional meaning. They are given scientific form and contribute to understandings of nationalism, identity, ethnicity and race. They are written about and published and legislated for and viewed and funded. Their DNA moves from being part of a living organism to being an archive of a dead thing, the echo of life, reanimated to mean again centuries later, haunting the now with a flicker of then-nowness. They become part of scientific discourse, circulating as data and argument; they are part of wider knowledge, signifying new approaches and the revelation of science; they are part of the museum, reified and made precise. The reduction of these in-dividuals to their individual genetics gives them agency and the ability to transform the world around them. By turning them into data it effects a change on their material. DNA analysis here is a way of seeing the past and making it live again, imaginatively, and textually. The imaginative tools that are being used to render this stuff animate, insofar as it gains agency and purpose and meaning, are those of the geneticist. The genetic profile of 'Cheddar Man' writes him a new identity, locate him differently, recreate his actuality. His genetic identity also allows a newly revisionist way of thinking about the past to be articulated, and he un-wittingly participates in a contemporary debate around the nature of English national identity and the relation between humans in the contemporary moment with the ancient past.

Decolonise DNA

Concluding this discussion of race, genetics, and politics, building on this chapter and looking to the next one regarding Ethics, this final section considers ways that structures of power and knowing, articulated through the use and interpretation of genetic data, might be resisted. Responding to James Watson's racism Sylvia Wynter argues that his thoughts represent 'the biocentric Scholasticism [...] of our present episteme'.[91] She explains:

> This is an episteme that functions, with respect to the knowledge of our contemporary world and its systematic reality, *according to the same cognitively closed descriptive statement and its sociogenically encoded truth of solidarity as that of the theo-Scholastic knowledge system of the medieval order of Latin-Christian Europe* (p. 20)

In contrast, Wynter argues, 'we cannot allow ourselves to continue thinking in this way [...] This is the enacting of a uniquely secular liberal monohumanist conception of the human' (pp. 20-21). Genetic history has the potential for challenging this model. As Linda Tuhiwai Smith argues, Indigenous voices challenge the fundamental assumptions and epistemologies of scholarship. She reminds us that: 'Indigenous peoples across the world have other stories to tell which not only question the assumed nature of those ideals and the practices that they generate, but also serve to tell an alternative story: the history of Western research through the eyes of the colonized'.[92] Yet Indigenous and native con-ceptions of history, time, and identity are often distinct and rarely taken into account in mainstream scientific, archaeological, and historical thought.[93] Ancient

DNA investigation led by native or Indigenous communities might assert different models of historical process. Such work has the potential to ensure that subaltern and non-Western voices are heard and ensure that dominant institutional and intellectual structures can be challenged. Ancient DNA work can challenge the centrality of whiteness to definitions of humanness and complicate notions of identity, race, and kinship.[94] Such work allows Indigenous peoples to develop links to longer histories than simply post-colonial or post-contact. This is a powerful tool in presenting alternative and resistant narratives for Indigenous and native peoples. It is also a moment when genetic information and insight becomes 'history', that is, evidence presented to challenge or revise a dominant narrative.

Information derived from Ancient DNA can complicate historical narratives, or present evidence that will enable modern-day marginalised communities to understand their historical development in more depth than had hitherto been allowed by 'mainstream' methods. Data can be used to present information about enslaved peoples and to work to give them a voice and some status in the historical record.[95] Those peoples who were erased from history are reconsidered, no longer marginalised from historical understanding.[96] Such analysis may provide the means for contemporary populations to understand links to marginalised or even extinct cultures.[97] This enables a challenge to extinction narratives, something that many Indigenous peoples have been undertaking for decades. Extinction narratives have been used to defend seizure of land and slow the recognition of Indigenous rights.[98] Ancient DNA analysis can challenge and subvert the narratives imposed by colonial settlers, supporting a challenge to normative historical discourse. For example, in Puerto Rico work on Ancient DNA has demonstrated the complexity of modern populations and allow for a challenge to settler colonial accounts of population development. At present the ways in which the 'native Indigenous communities resisted, survived, and were transformed [postcontact] is unclear. Their biocultural connection to present-day islanders is also disputed'.[99] Extinction narratives, 'based largely on colonial era censuses' are 'strongly opposed by islanders who assert cultural affiliation and direct descendance from native precontact communities' (p. 612). In response, work on genetic complexity can lead to a 'critical and interdisciplinary reassessment of historical narratives of Indigenous extinction' (p. 621).

As this example shows, genetic-led investigation can challenge the mainstream narratives presented by science and history, as well as enfranchise contemporary Indigenous peoples and suggest an unknown, hitherto unseen history for them. This type of work is explicitly a revisionist intervention, melding the scientific and activist historical approaches of decolonised research and 'informing future study of Indigenous responses to European colonization' (p. 621). Ancient DNA work can give insight into 'precontact genetic diversity' (p. 612), and demonstrate the range of originary peoples; it can enable a deeper and more fine-grained understanding of the ways that Indigenous peoples have been marginalised and disempowered.[100] Similarly work on ancient DNA can demonstrate development of new Indigenous practices under colonial rule, complicating the colonial history.[101]

However, there is much work to do to ensure that genetic knowledge does not simply commodify and develop new modes of control and manipulation.[102] The issue of decolonisation is real and not simply intellectual. It demands an activist historical intervention rather than straightforward reordering of priorities.[103] If as a practice it is to have effect then the practices and approaches of genetic history need to participate in the decolonisation project actively, to challenge and change the structures that exist and to expose the ways that peoples have been marginalised and disempowered. Specifically, the developments allowed by genetic investigation need to contribute to the wider decentring of certain narratives and modes of knowledge, not least those associated with normatively constructed 'history' itself. Decolonisation is a way of approaching the collection of objects, the power relations inherent in structures, dominant narratives, and the continuing effacement of peoples and voices.[104] In many ways it presupposes an ethics of engagement with the other that subverts contemporary ethics and epistemology, and in itself this seeks to decentre and reconfigure knowledge. As Dipesh Chakrabarty argues,

> There is a peculiar way in which all these other histories tend to become variations on a master narrative that could be called 'the history of Europe'. In this sense, 'Indian' history itself is in a position of subalternity; one can only articulate subaltern subject positions in the name of this history[105]

Decolonising DNA would involve recognising that the structures inherent in knowledge production about genetics are themselves part of the problem. Decolonising historical genetics might use DNA information to challenge normative structures, but also could seek to use DNA as a means for undermining or critiquing hierarchies of knowledge. As such, DNA would be a means for revealing the epistemologies that have been the foundation for particular notions of past, identity, race, and humanness, and to fracture them into something else. It would also involve giving voice to those marginalised, changing the story, and developing new models for knowing. Decolonising DNA as a mode, approach, or critique would recognise genetic science as having historical agency insofar as it has participated in the oppression and marginalisation of peoples and contributes to wider structural inequality. To 'decolonise DNA' as a means for restitution would involve genetic science contributing a new way of thinking about the present and the past, an interrogation of the normative. As Giles Deleuze and Felix Guattari argue, this might lead to a completely new way of knowing the human in the past:

> History is always written from the sedentary point of view and in the name of a unitary State apparatus, at least a possible one, even when the topic is nomads. What is lacking is Nomadology, the opposite of a history[106]

Decolonising DNA would involve challenging the current institutions that create, interpret and teach genetic knowledge and data. It would consist of questioning the correlation of biopower with political power. It involves a reflection upon the racist histories and practices that genetic science is predicated upon, and a reckoning with this beyond diversifying curricula and changing the names of laboratories. Decolonising DNA would challenge the equalling of improved western health with global progress.[107] Decolonising DNA would mean recognising the racist ideologies that most scientific powerbases are built upon, and the inequities of access to knowledge. It would involve recognising that post-Enlightenment epistemologies have created modes of understanding and constructing the world. It would mean challenging the centrality of modes of temporal interpretation, and the centrality of a type of scientific objectivity. Decolonising DNA would mean considering inequality and lack of voice, enabling the marginal; it would mean reordering and challenging. It would involve recognising that our contemporary situation is predicated upon colonial and imperial violence. It would involve restitution and reorganisation which is a politically active approach and, furthermore, a necessity. As Linda Tuhiwai Smith outlines,

> A critical aspect of the struggle for self-determination has involved questions relating to our history as Indigenous peoples and a critique of how we, as the Other, have been represented or excluded from various accounts [...] Indigenous peoples want to tell our stories, write our own versions, in our own ways, for our own purposes. It is not simply about giving an oral account or a genealogical naming of the land and the events which raged over it, but a very powerful need to give testimony to and restore a spirit, to bring back into existence a world fragmented and dying. The sense of history conveyed by these approaches is not the same thing as the discipline of history, and so our accounts collide, crash into each other.[108]

Notes

1 Paul Rincon, 'Cheddar Man: DNA shows early Briton had dark skin', *BBC News*, 23 February 2018, https://www.bbc.co.uk/news/science-environment-42939192 [accessed 9 August 2021].

2 Jenny Reardon and Kim TallBear, '"Your DNA is Our History": Genomics, Anthropology, and the Construction of Whiteness as Property', *Current Anthropology* 53: S12 (2012), S233–45. See also Amade M'Charek, 'The Mitochondrial Eve of Modern Genetics: Of Peoples and Genomes, or the Routinizaation of Race', *Science as Culture* 14:2 (2005), 161–83.

3 Keolu Fox, Kartik Lakshmi Rallapalli and Alexis C. Komor, 'Rewriting Human History and Empowering Indigenous Communities with Genome Editing Tools', *Genes* 11:1 (2020), https://doi.org/10.3390/genes11010088. In the following section I do not use the term 'Native American' apart from in citations as this is not a term used by communities to identify themselves. I recognise that not all of the work that I cite here is by Indigenous or native scholars but have attempted to use as much as possible, and I have sought to use publications that have worked in partnership with these communities.

4 Sarah Abel and Krystal Tsosie, 'Family History and the Global Politics of DNA', *International Public History* 2 (2019), https://doi.org/10.1515/iph-2019-0015, p. 3. See

also Debra Harry, 'Biocolonialism and Indigenous Knowledge in United Nations Discourse', *Griffith Law Review* 20:3 (2011), 702–28 and Frank Kressing, 'Screening Indigenous People's Genes: The End of Racism, or Modern Bio-Imperialism' in *Biomapping Indigenous Peoples* ed. Susanne Berthier-Foglar, Sheila Collingwood-Whittick and Sandrine Tolazzi (Leiden: Brill, 2012), pp. 117–36.

5 L.L. Cavalli-Sforza, A.C. Wilson, L.R. Cantor, R.M. Cook-Deegan and M.-C King, 'Call for a Worldwide Survey of Human Genetic Diversity: A Vanishing Opportunity for the Human Genome Project', *Genomics* 11 (1991), 490-1 (p. 490). See Sommmer, *History Within,* pp. 302–31.

6 Bryony Onciul, *Museums, Heritage and Indigenous Voices: Decolonizing Engagement* (London and New York: Routldge, 2015) and Amy Lonetree, *Decolonizing Museums: Representing Native America in National and Tribal Museums* (Chapel Hill, NC: University of North Carolina Press, 2012).

7 *Genetic Geographics*, p. 86.

8 Jonathan Marks, *What it means to be 98% Chimpanzee: Ages, People, and their Genes* (Berkeley, CA: University of California Press, 2002). See also the outline given by Emma E. Kowal, 'Genetics and Indigenous Communities: Ethical Overviews' in *International Encyclopedia of the Social & Behavioral Sciences*, 2nd edition, Volume 9 (2015), pp. 962–68.

9 M'Charek, *The Human Genome Diversity Project*, p. 2.

10 Laurelyn Whitt, *Science, Colonialism, and Indigenous Peoples* (Cambridge: Cambridge University Press, 2014), pp. 105–33 and M'Charek, p. 13.

11 Michael Dodson and Robert Williamson, 'Indigenous Peoples and the morality of the Human Genome Diversity Project', *Journal of Medical Ethics* 25 (1999), 204–8 (p. 207).

12 Jenny Reardon, 'The Human Genome Diversity Project: A Case Study in Coproduction',*Social Studies of Science* 31:3 (2001), 357–88 (370).

13 https://hagsc.org/hgdp/, see also Howard M. Cann et al, 'A Human Genome Diversity Cell Line Panel', *Science* 296: 5566 (2002), 261–2.

14 'The Human Genome Diversity Project: past, present and future', *Nature Reviews Genetics* 6 (2005), 333–340.

15 Adam Rutherford, *How to argue with a racist.*

16 Whitt, *Science, Colonialism, and Indigenous Peoples*, p. 1.

17 Emma Kowal, 'Disturbing Pasts and Promising Futures: The Politics of Indigenous Genetic Research in Australia', in *Biomapping Indigenous Peoples* ed. Susanne Berthier-Foglar, Sheila Collingwood-Whittick and Sandrine Tolazzi (Leiden: Brill, 2012), pp. 329–47.

18 Amade M'Charek, 'Beyond Fact or Fiction: On the Materiality of Race in Practice', *Cultural Anthropology* 28:3 (2013), 420–42 (p.

19 Jennifer Couzin-Frankel, 'DNA returned to tribe, raising questions about consent', *Science*, 328: 5978 (2010), 558.

20 Katherine Drabiak-Syed, 'Lessons from Havasupai Tribe v. Arizona State Board of Regents', *Journal of Health and Biomedical Law* 6 (2010), 175–225.

21 Nanibaa' A. Garrison, 'Genomic justice for Native Americans: Impact of the Havasupai case on genetic research', *Science, Technology and Human Values* 38:2 (2013), 201–23.

22 Robyn L. Sterling, 'Genetic Research amongst the Havasupai: A Cautionary Tale', *Virtual Mentor: American Medical Association Journal of Ethics* 13:2 (2011), 113–17. Rex Dalton outlined the legal and ethical complexities of the case at its inception in the charmingly titled 'When two tribes go to war', *Nature* 430 (2004), 500–02.

23 http://www.nature.com/nature/journal/v506/n7487/full/nature13025.html.

24 Jennifer A. Raff and Deborah A. Bolnick, 'Paleogenomics: Genetic Roots of the First Americans', *Nature*, 506 (2014), 162–3 (162).

25 Morten Rasmussen et al, 'The genome of a Late Pleistocene Human from a Clovis burial site in Montana', *Nature* 506 (2014), 225–9.

26 Cited in http://billingsgazette.com/news/state-and-regional/montana/remains-of-ancient-child-ceremoniously-reburied/article_3fcc174d-6f01-55b9-9923-96c9223ecda8.html

27 Morten Rasmussen et al, 'The ancestry and affiliations of Kennewick Man', *Nature* 523 (2015), 455–8.

28 Eske Willerslev outlines how his attitudes have 'evolved' in Carl Zimmer, 'Eske Willerslev is Rewriting History with DNA', *New York Times*, 16 May 2016, https://www.nytimes.com/2016/05/17/science/eske-willerslev-ancient-dna-scientist.html [accessed 2 April 2020].

29 Keolu Fox and John Hawks, 'Use ancient remains more wisely', *Nature* 572 (2019), 581–3 (582).

30 Maui Hudson et al, 'Rights, interests and expectations: Indigenous perspectives on unrestricted access to genomic data', *Nature Reviews Genetics* (2020), DOI: https://doi.org/10.1038/s41576-020-0228-x, my italics.

31 Maile Arvin, *Possessing Polynesians: The Science of Settler Colonial Whiteness in Hawai'i and Oceania* (Durham, NC: Duke University Press, 2019).

32 Jenny Reardon and Kim TallBear, '"Your DNA is *Our* History": Genomics, Anthropology, and the Construction of Whiteness as Property', *Current Anthropology* 53:S5 (2012), S233–45 (p. S235).

33 Peter A. Chow-White and Troy Duster, 'Do Health and Forensic DNA Databases increase Racial Disparities?', *PLoS Medicine*, 8:10 (2011), doi: 10.1371/journal.pmed.1001100.

34 Krystal S. Tsosie, Joseph M. Yracheta and Donna Dickenson, 'Overvaluing individual consent ignores risks to tribal participants', *Nature Reviews Genetics* 20 (2019), 497–8 (p. 498).

35 Kim TallBear, *Native American DNA: Tribal Belonging and the False Promise of Genetic Science* (Minneapoli, MN: University of Minnesota Press, 2013). See also P.A. Cochran et al, 'Indigenous ways of knowing: implications for participatory research and community', *American Journal of Public Health* 98:1 (2008), 22–7.

36 'mSphere of Influence: an Inupiat journey into science', *mSphere* 4:5 (2019), DOI: 10.1128/mSphere.00595-19.

37 Laura Arbour and Doris Cook, 'DNA on loan: issues to consider when carrying out genetic research with aboriginal families and communities', *Community Genetics* 9:3 (2006), 153–60, Gabriela Manaya and Joel Roque, 'Ethical problems in health research with Indigenous or originary peoples in Peru', *Journal of Community Genetics* 6: 3 (2015), 201–06, and Maile Tauali`i et al, 'Native Hawaiian views on Biobanking', *Journal of Cancer Education* 29 (2014), 570–6.

38 Maui Hudson et al, 'Rights, interests and expectations: Indigenous perspectives on unrestricted access to genomic data', *Nature Reviews Genetics* 21 (2020), 377-84 (377).

39 Katrina G. Claw et al, 'A framework for enhancing ethical genomic research with Indigenous communities', *Nature Communications* 9 (2018), https://doi.org/10.1038/s41467-018-05188-3.

40 See the discussion in Ripan S. Malhi and Alyssa C. Bader, 'Engaging Native Americans in Genomics Research', *American Anthropologist* 117:4 (2015), 743–4 and Angela Beaton et al, 'Engaging Māori in biobanking and genomic research', *Genetics in Medicine* 19 (2017), 345–51 and also Nicole K. Taniguchi, Maile Taualii and Jay Maddock, 'A comparative analysis of Indigenous research guidelines to inform genomic research in Indigenous communities', *The International Indigenous Policy Journal* 3:1 (2012).

41 Tri-Council Policy Statement: Ethical Conduct for Research Involving Humans – TCPS 2 (2018), in particular chapter 9, 'Research Involving the First Nations, Inuit and Métis Peoples of Canada', available through Panel on Research Ethics website, https://ethics.gc.ca/eng/policy-politique_tcps2-eptc2_2018.html [accessed 24 June 2021].

42 Deborah A. Bolnick, Jennifer A. Raff, Lauren C. Springs, Austin W. Reynolds and Aida T. Miró-Herrans, 'Native American Genomics and Population Histories', *Annual Review of Anthropology* 45 (2016), 319–40. See Rosalina James et al, 'Exploring pathways to trust: a tribal perspective on data sharing', *Genetics in Medicine* 16 (2014), 820–6.

43 Latrice Landry, Nadya Ali, David R. Williams, Heidi L. Rehm, Vence L. Bonham, 'Lack of Diversity in Genomic Databases is a Barrier to Translating Precision Medicine Research into Practice', *Health Affairs*, 37:5 (2018), 780–5.

44 Gillian M. Belbin, Maria A. Nieves-Colón, Eimear E. Kenny, Andres Moreno-Estrada and Christopher R. Gignoux, 'Genetic Diversty in Populations across Latin America: implications for population and medical genetic studies', *Current Opinions in Genetics & Development* 53 (2018), 98–104 (p. 99); see for instance Cesar Fortes-Lima et al, 'Exploring Cuba's population structure and demographic history using genome-wide data', *Nature Scientific Reports* 8 (2018), https://doi.org/10.1038/s41598-018-29851-3.

45 Darryl Leroux, *Distorted Descent: White Claims to Indigenous Identity* (Winnipeg, MB: University of Manitoba Press, 2019).

46 Hina Walajahi, David R. Wilson and Sara Chandros Hull, 'Constructing identities: the implications of DTC ancestry testing for tribal communities', *Genetics in Medicine* 21 (2019), 1744–50.

47 Walajahi, Wilson and Hull, p. 1746.

48 Deborah A. Bolnick et al, 'The Science and the Business of Genetic Ancestry Testing', *Science* 5849:318 (2007), 399–400 (p. 399).

49 Daryl Leroux, '"We've been here for 2,000 years": White settlers, Native American DNA and the phenomenon of indigenization', *Social Studies of Science* 48:1 (2018), 80–100.

50 Interviewed in Linda Geddes, 'There is no DNA test to prove you're Native American', *New Scientist*, 5 February 2014, https://www.newscientist.com/article/mg22129554-400-there-is-no-dna-test-to-prove-youre-native-american/ [accessed 2 April 2020].

51 Kim TallBear, *Native American DNA* (Minnesota MN, 2013), introduction, paragraph 6, Kindle edition.

52 Andrea Crossan, "You Took a DNA Test and It Says You Are Native American. So What?," *PRI's The World*, 24 November 2016, https://www.pri.org/stories/2016-11-24/you-took-dna-test-and-it-says-you-are-native-american-so-what [accessed 6 April 2020].

53 Reuters news service, 'Elizabeth Warren: DNA test finds 'strong evidence' of Native American blood', BBC online, 15 October 2018, https://www.bbc.co.uk/news/world-us-canada-45866168 [accessed 2 April 2020].

54 Annie Linskey, 'Elizabeth Warren releases results of DNA test', *Boston Globe*, 15 October 2018, https://www.bostonglobe.com/news/politics/2018/10/15/warren-addresses-native-american-issue/YEUaGzsefB0gPBe2AbmSVO/story.html [accessed 2 April 2020]. The original report has been deleted from its original website and therefore I am dependent on quoted details from early newspaper coverage. The *Boston Globe* broke the story in 2018 and was subsequently the basis for most global coverage.

55 Annie Linskey, 'Elizabeth Warren releases results of DNA test'.

56 Cherokee Nation (@CherokeeNation) Cherokee Nation responds to release of Senator Warren's DNA test. Image attachment. 15 October 2018.

57 Interviewed by Steve Inskeep in 'Determining who is a Cherokee is more than DNA, Hoskin says', *NPR* 16 October 2018, https://www.npr.org/2018/10/16/657749867/determining-who-is-a-cherokee-is-more-than-dna-hoskin-says [accessed 6 April 2020].

58 Jennifer Raff, 'What do Elizabeth Warren's DNA Test Results Actually Mean?', *Forbes* 15 October 2018, https://www.forbes.com/sites/jenniferraff/2018/10/15/what-do-elizabeth-warrens-dna-test-results-actually-mean/#51e9cb0312df [accessed 6 April 2020].

59 Kim TallBear (@KimTallBear) 'after too many media inquiries, here is my statement on the #ElizabethWarren DNA testing story'. Image attachment. 15 October 2018.

60 Asma Khalid, 'Warren Apologizes to Cherokee Nation for DNA test', *NPR* 1 February 2019, https://www.npr.org/2019/02/01/690806434/warren-apologizes-to-cherokee-nation-for-dna-test?t=1585841469756 [accessed 6 April 2020].

61 See Leroux, *Distorted Descent*.

62 Nicholas W. Gillam, 'Sir Francis Galton and the Birth of Eugenics', *Annual Review of Ethics* 35 (2001), 83-101; Josh Gabbatiss, 'DNA pioneer James Watson has final honours stripped amid racism row', *The Independent*, 12 January 2019, https://www.independent.co.uk/news/science/james-watson-racism-honours-dna-double-helix-cold-spring-harbor-laboratory-a8724896.html [accessed 15 May 2020].

63 Surveyed in Adam Rutherford, *How to Argue with a Racist* (London: Weidenfeld & Nicolson, 2020).

64 Rutherford, *How to Argue with a Racist*, Introduction, paragraph 6, Kindle edition.

65 Aaron Panofsky and Joan Donovan, 'Genetic ancestry testing among white nationalists', *Social Studies of Science* 49:5 (2019), 653–81.

66 Sarah Zhang, 'Will the Alt-Right Promote a New Kind of Racist Genetics?', *The Atlantic* 29 December 2016, https://www.theatlantic.com/science/archive/2016/12/genetics-race-ancestry-tests/510962/ [accessed 15 May 2020].

67 Simon Outram et al, 'Genes, Race, and Causation: US Public Perspectives about Racial Difference', *Race and Social Problems* 10 (2018), 79–90.

68 Explored in Reardon, *Race to the Finish* and Nelson, *The Social Life of DNA*.

69 *Tribes* (London: Little, Brown, 2020).

70 Volker Heyd, 'Kossinna's smile', *Antiquity* 91 (2017), 348–59.

71 'On the biodeterministic imagination', *Archaeological Dialogues* 27 (2020), 1–16.

72 Jenny Reardon, 'The Democratic, Anti-Racist Genome? Technoscience at the Limits of Liberalism', *Science as Culture* 21:2 (2012), 25–47 (p. 26).

73 Discussed in Jerome de Groot, 'Ancestry.com and the evolving nature of historical information companies', *The Public Historian*

74 Vineet Mehra, Ancestry Chief Marketing Officer, quoted in press release 'In Honor of Fourth of July, Ancestry Honors the Past and Celebrates the Future', https://www.ancestry.com/corporate/newsroom/press-releases/honor-fourth-july-ancestry-honors-past-and-celebrates-future [accessed 24 January 2018].

75 Brad Argent quoted in Kristen Hyde, 'How British Are You?', Ancestry.com blog, 13 September 2016, https://blogs.ancestry.com/ancestry/2016/09/19/how-british-are-you-dna-study-reveals-uks-ethnic-diversity/ [accessed 13 January 2017].

76 Kristen Hyde, 'How British Are You?'.

77 Catherine Nash, *Genetic Geographies: The Trouble with Ancestry* (Minneapolis, MN: University of Minnesota Press, 2015).

78 Described in C.G. Seligman and F.G. Parsons, 'The Cheddar Man: A Skeleton of Late Paleolithic Date', *The Journal of the Royal Anthropological Institute of Great Britain and Ireland* 44 (1914), 241–63.

79 Reported in Selina Brace et al, 'Ancient genomes indicate population replacement in Early Neolithic Britain', *Nature Ecology & Evolution*, 3 (2019), 765–771.

80 Ceylan Yeginsu and Carl Zimmer, 'Cheddar Man, Britain's oldest skeleton, had dark skin, DNA shows', *New York Times*, 7 February 2018, https://www.nytimes.com/2018/02/07/world/europe/uk-cheddar-man-skeleton-skin.html [accessed 6 August 2021].

81 Iñigo Olalde et al, 'Derived immune and ancestral pigmentation alleles in a 7,000-year-old Mesolithic European', *Nature* 507 (2014), 225–8.

82 Kenneth Brophy, 'The Brexit hypothesis and prehistory', *Antiquity* 92: 366 (2018), 1650–8.

83 'Today, virtually no analysis is done on human remains without proposals that must pass through and be approved by many different groups. Prior to any analyses,

consensus and some form of cooperative effort between bioarchaeologists and other stakeholders must be obtained [...] The product of this more collaborative and consensual effort is not only a much deeper engagement with descendant communities but also a more detailed understanding of the human remains themselves', Debra L. Martin, Ryan P. Harrod, Ventura R. Perez, *Bioarcheology: An Integrated Approach to Working with Human Remains* (New York, NY: Springer, 2013) p. 4.

84 Thomas Laqueur, *The Work of the Dead* (Princeton, NJ: Princeton UP, 2015).

85 This is not unusual in archaeogenetics, see Catherine J. Frieman and Daniela Hofmann, 'Present pasts in the archaeology of genetics, identity, and migration in Europe: a critical essay', *World Archaeology* 51 (2019), 528–45.

86 Holly Christodolou, 'Who was Cheddar Man?', *The Sun*, 18 February 2018, https://www.thesun.co.uk/tech/5518817/cheddar-man-first-modern-brit-dark-black-skin-blue-eyes/ [accessed 6 August 2021].

87 Colin Fernandez, 'Face of the first Briton is revealed', *Daily Mail*, 7 February 2018, https://www.dailymail.co.uk/sciencetech/article-5358699/First-Brit-dark-skinned-blue-eyed.html [accessed 6 August 2021].

88 Comments at https://www.dailymail.co.uk/sciencetech/article-5358699/First-Brit-dark-skinned-blue-eyed.html [accessed 6 August 2021].

89 Comments at https://www.dailymail.co.uk/sciencetech/article-5358699/First-Brit-dark-skinned-blue-eyed.html [accessed 6 August 2021].

90 Comments at https://www.dailymail.co.uk/sciencetech/article-5358699/First-Brit-dark-skinned-blue-eyed.html [accessed 6 August 2021].

91 Sylvia Wynter and Katherine McKittrick, "Unparalleled Catastrophe for Our Species? Or, to Give Humanness a Different Future: Conversations" in *Sylvia Wynter: On Being Human As Praxis*, ed. Katherine McKittrick (Durham, NC: Duke University Press, 2014), 9–90 (p. 20).

92 Linda Tuhiwai Smith, *Decolonizing Methodologies* (London: Zed Books, 2012), p. 2. See also the essays in Devon Abbott Mihesuah and Angela Cavender Wilson, eds., *Indigenizing the Academy: Transforming Scholarship and Empowering Communities* (Lincoln, NE: University of Nebraska Press, 2004).

93 Leo Killsback 'Indigenous Perceptions of Time: Decolonizing Theory, World History, and the Fates of Human Societies', *American Indian Culture and Research Journal* 37:4 (2013), 85–114 and Angela Cavender Wilson, 'Reclaiming our humanity: Decolonization and the Recovery of Indigenous Knowledge' in *War and Border Crossings* ed. Peter A. French and Jason A. Short (Oxford: Rowman and Littlefield, 2005), pp. 255–65. See also Alexis Wright, *The Swan Book* (London: Constable, 2015). See the discussion in Marnie Hughes-Warrington, *Big and Little Histories: Sizing up Ethics in Historiography* (London and New York: Routledge, 2021).

94 Amanda Behm, Christienna Fryar, Emma Hunter, Elisabeth Leake, Su Lin Lewis and Sarah Miller-Davenport, 'History on the Line: Decolonizing History: Enquiry and Practice', *History Workshop Journal* 89 (2020), 169–91.

95 Rodrigo Barquera et al, 'Origin and Health Status of First-Generation Africans from Early Colonial Mexico', *Current Biology* 2020, DOI: https://doi.org/10.1016/j.cub.2020.04.002.

96 Hannes Schroeder et al, 'Genome-wide ancestry of 17th-century enslaved Africans from the Caribbean', *PNAS* 112 (2015), 3669–73; Anuradha Jagadeesan et al, 'Reconstructing an African haploid genome from the 18th century', *Nature Genetics* 50 (2018), 199–205.

97 Juan Carlos Martinez-Crusado et al, 'Mithchondrial DNA analysis reveals substantial Native American ancestry in Puerto Rico', *Human Biology* 73:4 (2001), 491–51.

98 Maximilian C. Forte, 'Who is an Indian? The Cultural Politics of a Bad Question', in *Who is an Indian?: Race, Place, and the Politics of Indigeneity in the Americas* ed. Maximilian C. Forte (Toronto: University of Toronto Press, 2013), 3–52.

99   Maria A Nieves-Colón et al, 'Ancient DNA Reconstructs the Genetic Legacies of Precontact Puerto Rican Communities', *Molecular Biology and Evolution* 37:3 (2020), 611–26 (p. 612).

100 See also Lars Fehren-Schmitz et al, 'Genetic Ancestry of Rapanui before and after European Contact', *Current Biology* xxvii (2017), 3209–15.

101 Tiffiny A. Tung et al, 'Constrained Agency while Negotiating Spanish Colonialism', *Bioarchaeology International* 3:3 (2019), DOI: http://dx.doi.org/10.5744/bi.2019.1013.

102 See Keolu Fox, 'The Illusion of Inclusion – the 'All of Us' Research Program and Indigenous People's DNA', *The New England Journal of Medicine* 383 (2020), 411–13 and Krystal S. Tsosie, Keolu Fox and Joseph Yracheta, 'Genomics data: the broken promise is to Indigenous people', *Nature* 591: 529 (2021), doi: https://doi.org/10. 1038/d41586-021-00758-w.

103 Eve Tuck and K. Wayne Yang, 'Decolonization is not a metaphor', *Decolonization* 1:1 (2012), 1-40 and Kirisitina Sailiata, 'Decolonization' in *Native Studies Keywords* ed. Stephanie Nohelani Teves, Andrea Smith and Michelle H. Rahjeta (Tucson, AZ: University of Arizona Press, 2015), 301–08.

104 Amy Lonetree, *Decolonizing Museums: Representing Native America in National and Tribal Museums* (Chapel Hill, NC: University of North Carolina Press, 2012).

105 *Provincializing Europe*, p. 27.

106 *A Thousand Plateaus*, trans. Brian Massumi (Minneapolis, MS and London: University of Minnesota Press, 1987), p. 23.

107 Whilst often considered a humanities concept, decolonising principles have been suggested in several branches of science, see Christopher H. Trisos, Jess Auerbach and Madhusadan Katti, 'Decoloniality and anti-oppressive practices for a more ethical ecology', *Nature Ecology & Evolution* (2021), https://doi.org/10.1038/s41559-021-01460-w and Robin Bronen and Patricia Cochran, 'Decolonize climate adaptation research', *Science* 372: 6548 (2021), p. 1245.

108 *Decolonizing Methodologies*, pp. 28–9.

4

ETHICS

Genetic data seems to present new knowledge about the past, but there are multiple issues around the collection, interpretation, and use of this knowledge. If, as Anton Froeyman argues, ethics is at 'the heart of the discipline [history]', then it should follow that the use of DNA to think about and narrate the past would have an ethics associated with it, or have an ethics articulated around or by it (and would be different to bioethics).[1] This chapter considers the impact that new genetic information might have on an ethics of approaching, and narrating, the past.[2] I explore several moments that allow an investigation of genetic ethics in relation to historical investigation.[3] The phenomena described in this chapter present several ethical challenges for understanding the past using genetic information. Primarily, these centre around moral decisions regarding the usage and manipulation of information. This particularly relates to the ways in which DNA might be considered 'evidence' and how this might be used by historians, law enforcement, and the public. There are questions relating to the data itself, its collection, issues of privacy and archiving. However, once these 'material' ethical questions have been satisfied more are raised about the ethical aspect of the arrangement of genetic information as evidence, the ethics of narrating the past, and the representation of history through the use of this data. Finally, it is important to interrogate the notion of DNA as something transhistorical. In the debate about de-extinction we can see the compelling argument that DNA is historically contingent. To 'use' it outside of its specific moment is to engage in unethical practices that, as imagined in *Jurassic Park*, are incredibly destructive. DNA here is a continuum between then and now, but as with all historical narratives that link the past with the present, it is one that reveals the ethics of representation.

Evidently there are ongoing debates surrounding the ethics of DNA, ranging from the surveillance use of information by nation states to questions of ownership of genetic data (some of this was discussed in the previous chapter). Whilst there

DOI: 10.4324/9781003052975-5

are clearly significant debates ongoing about the use, interpretation, and collection of biogenetic information, there is a further ethical consideration needed for those interested in history. The key concerns for my discussion here relate to the ways in which data is used to engage with or create the past. The use of DNA information in law enforcement is one way in which this aspect of genetics works. DNA also enables the investigation of genocide through the analysis of human remains. In these examples DNA provides information about the past that is impossible to generate in other ways, demonstrating its key importance as a communicator of information. It is something private that is being used in a public context. It is a kind of evidence that allows an understanding of events in the past to become clearer. The intervention of genetic knowledge is part of an address to a kind of truth hitherto hidden. DNA can help to reveal and open up the past to us, providing a clarity and for many a closure.

It is important to think about the status of DNA as evidence, to consider the ways in which it witnesses the past, and to consider its manifestation in historicised discourse. In particular it is important to recognise the imbrication of objectivity and truth in the discourses connected to genetic history. Genetic science is popularly considered to tend toward a kind of truth, deriving not just from its status as scientific discourse but also its wide usage in contexts of evidence (criminal justice, paternity, identity). Hence the use of genetics in historical narratives or constructions imports a type of 'evidence' or 'witness' that is (largely) beyond question. As we have seen in archaeology, often the genetic data is considered to be a fixed point around which all other evidence must reorient itself (see Chapter 2). However DNA is material, that is, a data that we can 'read', 'see', and manipulate. The more sophisticated our ability to engage with this material, the more stories and narratives we can tell. We do, however, need to critique and audit the ways that we engage with and understand DNA as this obviously impinges upon the translation of the evidence into meaning. We should be able to undertake theoretical and critical analysis of DNA simply as a thing, submitting 'it' to a number of investigative processes and investigations in order to ensure the robustness of the evidence in our historical practice.

Part of the ethical aspect of DNA is a popular understanding of the impact genetics has on the contemporary moment. Discussions around ethnicity and health focus on the significance for the participant in the present of aspects of their genetic past. This is part of a wider sense of social 'learning' from the past, although this discourse is here unique to a particular set of people (present participant and their family; their ancestors). Rather than society understanding itself better through the moral action of engaging with the past in this case individuals and their families might comprehend more fully through 'reading' their ancestry. Thus, the popularly understood post-Enlightenment moral impetus of 'history' – that is, to 'teach' through appreciating the complexity of the past – is here mapped onto the individual. There is a certain moral aspect here, too, in the discourses of health particularly – if you *could* know this information, the implication is, why *would* you not (for yourself, but also your family)?

If DNA is a mode of knowing the past, though, does it become more ethically problematic? Genetic information relating to a particular Indigenous group, for instance, can be thought to define them, allowing them to be known (and manipulated), and constructing them. We need to think about how genetic 'evidence' might be defined in respect of ethics. What type of evidence is it? Has it been collected fairly, or might the circumstances of its collection (or the decisions made about *what* to collect, or where) be criticised? What do we do with historically collected materials that may have been gathered within different (or no) ethical frameworks? Does looking at the way that such databases have been constructed enable a critique of process? DNA information creates an archive, which should be subject to the kinds of interrogation that any body of data would receive. What does that archive mean, how and why has it been collected, what is missing and what do the overall principles of collection betray about those who created it and the ways in which they seek to use it? How might it be decolonised? Jacques Derrida reminds us that an archive is a '*place* from which *order* is given' and genetic information participates in this attempt to impose structure and order.[4] How is this information stored, and treated, and interpreted, and read? All of these are ethical issues and all impose a new dimension upon what is considered 'inert' information or data. The answers to these various questions impinge upon the 'agency' of the DNA and the way that it relates to the individual it was taken from.

An important example here would be to consider the use of DNA from burial sites in archaeological and historical investigation (see Chapter 2). Thomas Laqueur has recently written about the 'work' of the dead and the ways in which encounters with the dead structure our social and cultural world.[5] His interest in the 'ontological standing' (p. 18) of the dead is important in thinking about the status of seemingly inert data derived from DNA. How does such data relate to its 'original'? If it has been derived from bodies long dead, what are the ethics of using this information without permission? The point about the 'ontological standing' is problematic in the case of DNA information, as it lives outwith the original. Genetic information is *never* neutral and always has a link to its source material, the physical body it was taken from. In the gap between the physical and the data we can discern a whole ethics of articulation.[6] The body is forgotten in the new space of data, as the individual becomes interpreted according to the information they might reveal. Yet as asserted, the body is always there, ghostly, the spectre that has produced the information. Laqueur looks at the material ways in which the dead echo in the contemporary, and the relationship between then and now, and memorialisation, that is implied. He argues that

> The history of the work of the dead is a history of how they dwell in us – individually and communally. It is a history of how we imagine them to be, how they give meaning to our lives, how they structure public spaces, politics and time. It is a history of the imagination, a history of how we invest the dead – again, I will be speaking primarily of the dead body – with meaning (p. 17).

The ethical point accrues around how we 'invest' the dead with meaning. If we interpret, read, investigate and understand using information derived from the physical bodies of the dead we are making them speak again. This is profoundly important and highly resonant. Bodies long dead begin to have a relationship with the contemporary moment, to *mean* again.

Historical Violence and Genetic Evidence

Genetic information and forensic DNA testing is frequently used in inquiries relating to violent events from the past.[7] DNA evidence is being used, for instance, to identify the dead in mass graves from the 1914-18 war. DNA evidence in is brought to bear in investigation of specific violence against particular communities, or, more rarely, in instances of natural disasters. In particular, such data is commonly part of investigation into violence such as genocide, human rights abuses, or massacre.[8] The use of such genetic evidence, in addition to techniques from pathology, is part of what has been described as the 'forensic turn' in investigation of genocide and large-scale violence.[9] This 'forensic turn' describes a movement towards understanding and interpreting mass violence through precise investigative techniques. DNA here is part of a suite of tools being used to approach investigation, enabling the revelation of events and shedding light on what has been hidden.

In particular forensic DNA evidence from historical war crimes such as genocide has been used to build cases against perpetrators and to identify victims. DNA evidence has been prominently used in the former Yugoslavia, for instance, and also in Rwanda, Syria, Argentina, Guatemala, Chile, South Africa and Spain.[10] The International Commission on Missing Persons (ICMP) has worked in most of these areas with DNA and also in locations of major disasters. Genetic work here helps to identify remains and is a means for redressing historic crimes. Genetic investigation is fundamental in enabling communities to gain justice for events in the past.[11] In ethical terms genetic data is used to render justice for those denied it, and to give voice to the histories of marginalised and erased communities and individuals.

In Argentina, since the mid-1980s a movement has accrued around the DNA of those stolen children whose parents were murdered by the state. Lindsay Adams Smith outlines the story of Pichi, a son of one of the disappeared: 'He explained that in his memory and, equally importantly, in his body, he was a repository of these multiple missing, a living substitute for his devastated family tree'.[12] Through the intercession of genetic science communities can remember and reclaim. Forensic genetics has become a way of families to recreate: 'In Argentina, DNA has become a primary technical tool for reconciling the atrocities of the past' (p. 1054). DNA has similarly been important in Spain, both in identifying remains and in enabling a debate about commemoration; the first 'disappeared' person from the Civil War identified this way was during exhumations of mass graves in 2003.[13] Smith argues more widely that DNA work has become crucial in enabling communities to gain justice, considering genetic analysis a 'human rights technology' (although this aspect has diminished post-9/11).[14] DNA analysis of evidence is being used to reclaim justice for those dead and to gain closure for those still living. Genetic work here can make concrete the absences and spectres of the past. The stories that DNA evidence allow to be told are those that had seemed lost, or buried, or ignored.

For some communities the development of a DNA archive seems to allow the possibility of their reinsertion into 'history' and a challenge to the ways that the past has been constructed and narrated. Genetic information has enabled the identification of slaves in burial grounds in the USA.[15] It has also been instrumental in generating discussion about the commemoration of burial grounds across the nation.[16] The activist Deadria Farmer-Paellmann has used DNA tests in her work demanding reparations from those who profited from slavery.[17] In particular she used commercial DNA tests as part of class actions bought on the behalf of descendants of slaves in 2002 and 2004. Farmer-Paellmann therefore 'articulates a *genetic* kinship to a contemporary group as an attempt to make a legally cognizable connection'.[18] As Alondra Nelson has argued, this was one of the first times that 'an upper-level court has had to deal with commercial and cultural output of the human genome project'.[19] In these actions DNA testing became a way of connecting to places of origin and legally to bring a challenge related to historical crimes. The lawsuits did not succeed, but the wider premise – the link between descendants now and historical events – was clearly made to the public. As such, perhaps 'DNA can offer an avenue toward recognition, but cannot stand in for reconciliation' (Nelson Wailloo, p. 29).

Henry Louis Gates, Jr, a famous advocate of DNA testing for family history (see discussion in Chapter 6), points out the potential for this new technology to provide those forgotten or erased by state and nation archives to assert their historical agency: 'For the first time since the 17th century, we are able, symbolically at least, to reverse the Middle Passage'.[20] This extraordinary statement suggests that genetic work allows a reconnection with family and with community. He remembers that 'searching for my ancestry' had ever been 'a fraught process, always a mix of joy, frustration and outrage, as the reconstruction of their history – individually and collectively – must always be for any African American' (p. 5). In contrast he outlines the power of DNA work: 'when the paper trail would end, as it inevitably did, in the horrid darkness of slavery, we traced our African roots through our DNA' (p. 11). DNA work shows the often unseen violence inherent in archives which themselves can enshrine attitudes and practices through what is not collected, or what is erased.

In Gates, Jr's framing, the intervention of genetic evidence connects a community. This has historiographical consequence, as a more diverse and more complete picture of the past might be drawn using the additional data in the archive: 'Restoring the stories of the lives of the members of our extended families can directly transform the way that historians reassemble the larger narrative of the history of our people' (p. 12). In this reading genetic evidence allows the circumvention of the state archive, finding a way of writing forgotten peoples into history. For Gates, Jr, genetic investigation provides a new type of connection to the past, one that challenges the stories written by archives and instead seeks new ways of knowing and telling the past. The language used here is instructive – 'reconstruction', 'restoring', 'reassemble', 'reverse' – as Gates, Jr, articulates the impact of DNA as having a recuperative effect and contributing to the recreation of the past. Genetic data hence is seen here as parallel to mainstream evidence (it has similar effect) but also the action of restorative justice. Using the information in this way hence ensures that historical genetic data has an ethical aspect, insofar as it is part of the reconstitution of the contemporary self and part of a reclaiming of the past by particular communities.

The use of genetic information for reconciliation and reconstitution is increasingly common, and these projects allow discussion of institutional and structural racism, as well as revealing the power dynamics of the ways in which information is collected, kept, and disseminated.[21] As Yulia Egorova points out, however, whilst it is 'tempting' to see DNA work as something that 'subaltern communities can use as a tool of empowerment in projects of identity arbitration or reconciliation', very often these seemingly empowering projects can become a 'tool of subordination, marginalization or oppression'.[22] Simplifying information about communities, imposing narratives and misinterpreting data can do as much harm as good in terms of restitution. Many scholars and communities are deeply suspicious of genetic understanding of ethnicity, identity or heritage, and point out that to reduce 'race' to genetic and biological data can perpetuate and contribute to racist attitudes.[23] In particular the popularity of misunderstood or pseudo-scientific genetic definitions of race must be challenged.[24] As these critical scholars have pointed out, genetic data is extremely potent and has the potential to misread entire communities or warp the ways in which people think about their pasts (this is discussed further in Chapter 4).

The controversies over using DNA to interpret and understand the past demonstrate what is at stake with this new technology. Genetic data in these contexts is being used as evidence, to retell and restore narratives of the past, to shed light on old events, to restore justice, to commemorate and memorialise, and to intercede. The entry of genetic science into the legal and human rights discourse means that as a form of historical evidence linking to the present it becomes extremely powerful. The information has an ethical dimension in its reconciliatory action. Genetic information is being harnessed in order to understand and reorder the past, whether correctly or not. Indeed, the misuse and problematic interpretation of DNA data highlights the ethical problems inherent in reading and interpreting information and evidence. The question of who gets to 'speak' becomes increasingly important as it becomes clear that genetic information does not 'speak' blandly, or, rather, what it is made to say is never neutral. In many ways DNA data is simply revealed here to be as malleable and problematic as any type of evidence, despite its seeming novelty, power, and objectivity. Yet the influence of this evidence, its wider impact upon social attitudes and understanding, ensures that as evidence it 'means' more. It has an impact upon the ways in which memory is structured and understood. This is what Alondra Nelson terms the 'social life' of DNA, the fact that it is portable and malleable, moving and shifting in meaning and interpretation through multiple areas of society.[25]

Cold Cases

This section looks at the use of DNA as part of criminal investigation procedure, where the genetic evidence links the perpetrator to a crime or historical DNA is used to identify individuals. This is the most significant aspect of DNA in the popular imagination and contributes to the understanding of genetics as evidence.[26] DNA evidence links the perpetrator to a criminal event in the past, reaching to identify them in the present. Genetic information is a means for investigating what had formerly been unknown and presents evidence from the past. In doing so it opens up a new ethical discussion about the value of information, as well as reconfiguring notions of 'evidence' as a legal and historical category. Genetic genealogy is explored in detail in Chapter 6.

Genetic Genealogy and Crime

Genetic science has been used in crime investigation from the early 1980s, when DNA profiling ('fingerprinting') developed in the UK and the USA.[27] DNA fingerprinting 'revolutionized forensic investigations', providing entirely new lines of investigation.[28] There are important consequences of this for understanding of the citizen and a wider understanding of the way the body might be configured and used as a means of control. As a result of improved technology the 'legislation underpinning DNA sampling has reframed the human body in law and redrawn its intimate boundaries'.[29] Genetic identity becomes something enforceable, inescapable, and part of the state's work of interpellating the population. Writing in the *Observer* on the 25th anniversary of Alec Jeffreys's work on the technology, Robin McKie suggested that 'the DNA fingerprint has become part of a civic apparatus that can follow the movements of individuals with unprecedented accuracy'.[30] This sense that we are all trackable due to our DNA communicates its function as a surveillance technology. Whether we want to or not, we leave evidence of ourselves everywhere we go it seems. The idea that we are leaving a trail of DNA behind us that could be followed by the right investigation adds a frisson to everyday life, as well as reminding us that the bodily 'self' we might claim to be is constantly moulting and disintegrating. The plurality of DNA evidence, and its wide acceptance in popular culture (discussed below), suggests that populations are generally comfortable with being understood in this way.[31] However, this is a very modern understanding of the self in relation to criminality. During the O.J. Simpson trial in 1995 the prosecution team discovered that lack of understanding of DNA evidence hampered their case substantially.[32] In recent years, conversely, the use of DNA information in crime scene analysis and investigation is now so normalised that without such evidence juries are often tentative to convict.[33] The evidentiary impact of genetic analysis, therefore, is something that has only occurred in the past two decades.

The biggest law enforcement collection of genetic information in the world is the UK National DNA database, which has around 6.4 million sample profiles and $c.630000$ samples from crime scenes.[34] The database is over 80% male and 75% White European, covering around 9.5% of the UK population. This is an enormous archive of information which is applied to ongoing and cold cases.[35] There are wide-ranging ethical issues relating to privacy and usage of DNA in criminal cases, and to the storage of such information. For our purposes it is important to note that DNA information is used as evidence, part of the wider narrative of the crime. Genetic specificity, and the ability to locate genetic material, identifies and articulates the particularity of the individual within the wider population. It allows criminality to be revealed and contributes to law enforcement.

Genetic genealogy resources have been used regularly to identify remains (see above).[36] Professional and volunteer genealogists have worked with the police in identification of bodies that have been unknown for decades.[37] This process involves comparing DNA results with various genetic databases, in particular the GEDMatch database. GEDMatch is a free community-created database that allows users to upload genetic information from the major commercial databases. It has a number of tools for manipulation of that data. The DNA Doe project is the organisation that first used GEDMatch DNA to identify human remains. This project works to use genetic genealogy to identify unknown bodies. The researchers in the project use GEDmatch as a research tool and then, if there is a match, build trees to locate contemporary family of missing persons. They have been relatively successful and have become increasingly well-used by law enforcement organisations. Whilst they are professional genealogists, their project is dependent on donations to survive.

However, increasingly genetic genealogy is being used by police groups as a means of breaking cold cases rather than simply identifying remains. In April 2018 police in California announced that they had arrested a suspect in the so-called 'Golden State Killer' case. The man was accused of being the 'Golden State Killer' who murdered and sexually abused women in the 1970s and 1980s. The case had been considered 'cold' for some decades before police decided to use the genetic genealogy site GEDMatch. The police uploaded a DNA profile taken from one of the crime scenes, and generated multiple matches to possible family members. They used this set of possibilities to begin to establish a link to the criminal. Once they had a suspect they used more DNA tests to ensure that their match was correct. This led to the arrest of a man and the case coming to court. It has been estimated that around 25 cold cases have been solved in this way, with the 'Golden State Killer' the most high profile at present.[38] William Talbott II was the first person convicted in a case using genetic genealogy.[39] The termination of the case gave justice to the victims and helped their families to some kind of closure.

The genealogy work is treated as a tip, part of the investigation that (in the Talbott case) is not challenged by defence lawyers. Genetic genealogy is being used most often in cases relating to violent crime. However it is also being used as an approach in Canada and the USA to identifying the mothers of abandoned babies, with several cases pending.

Genetic genealogy at present stands as the confluence of two types of approach to evidence and genetics: one corporate, the other more volunteer-led. The use of community databases in the investigation of crime is challenging to old conceptions of privacy. Thus on the one hand is the DNA Doe project, run by volunteers, and on the other is the Parabon Nanolabs company, keen to demonstrate their trademarked motto 'Get More From Your DNA Evidence'.[40] Parabon have a service entitled 'Snapshot' which combines Genetic Genealogy with other DNA-related predictions such as Phenotyping. As their marketing suggests, 'Snapshot is ideal for generating investigative leads, narrowing suspect lists, and solving human remains cases, without wasting time and money chasing false leads'.[41] Parabon are working hundreds of cases across the United States and their successes in identifying perpetrators were up to 69 in 2019.[42]

Indeed, GEDMatch became, for a while, a key resource for law enforcement. Accordingly, GEDMatch changed its privacy conditions in June 2019 and required users to opt-in for their information to be used in this way; FamilyTreeDNA, who had formerly worked with the FBI, also changed their privacy settings after apologising for this in 2019.[43] Immediately the resource open to investigation was significantly reduced as many users removed themselves. However, the concept of using DNA in this way remained interesting to law enforcement and in particular to border patrol. Like many other tech companies, genetic genealogy organisations faced pressure to enable them to use their information.

The use of genetic data aligned with family history information is known as a 'long-range familial search'.[44] In 2018 around 13 cases were 'reportedly solved by long-range cases' but there is an increase in using them in 'active investigations'; the approach looks like becoming a 'standard investigative tool'.[45] The approach is incredibly powerful. The number of people reachable by such work is enormous, so from a relatively small sample (GEDMatch had around 1 million users when it was first being interrogated as a dataset) it is possible to generate matches for many millions more. Some work has demonstrated broadly 'a genetic database needs to cover only 2% of the target population to provide a third-cousin match to nearly any person' and hence' the technique could implicate nearly any U.S. individual of European descent in the near future'.[46] Many have expressed alarm at the 'appropriateness of a near-national DNA database by happenstance, rather than by democratic deliberation and legislation'.[47] The fact that the databases are relatively public also means that the uploading of a suspect's DNA 'violates the suspect's constitutional right to privacy' as their information is being shared with a huge number of people.[48] Critics argue that such 'law enforcement methods of genetic investigation are both haphazard and underregulated'.[49]

This has implications for privacy and introduces a new dimension into genetic understanding of the past: the law. On the one hand, this is a utilitarian way of approaching genetic genealogocal information as a simple database that can enable a new avenue of investigation to be opened up. Law enforcement is simply identifying ways of 'reading' the data that had not been considered beforehand. On the other hand, it is directly using the tools of genetic genealogy as a kind of shortcut to break cases without consideration for the multiple people who will become involved in a murder investigation without their knowing, and the multiple privacy and procedural issues that this investigation entails.[50] There is an ethics developing around the body as a repository of different historical information: individual, species, legal.

The questions of privacy surrounding genetic data online are complex. Medical, legal and research databases are highly regulated. This is not the case necessarily for more commercial use. Whilst there is some oversight on Genetic Testing Direct To Consumer firms the increasing complication and diversification of the market means that there is very little regulation of 'entities whose services consist solely of interpreting, reinterpreting, or facilitating self-interpretation of individuals' raw genetic data'.[51] Hence whilst there may be strong ethical and privacy policies for some of the work being done there is less in the newer providers or the community groups. This is demonstrated by the shifts in privacy policy that many of the organisations have undertaken after the high profile criminal cases.[52] As legal scholars have pointed out, the contracts that are entered into with online services are often highly binding and little understood.[53]

The moral dimensions of such use of genetic genealogy are manifold. The use of information collected for one purpose becoming part of a criminal investigation tests contemporary notions of privacy.[54] Many users seem comfortable with their data being used in the investigation of murder and violent crime, despite the fact that they were not originally consulted about this. Yet there are clear concerns that this approach might eventually be used more widely.[55] The moral issue becomes more muddied when considering the use of genetic genealogy being used by border patrol, for instance, or in wider profiling rather than particular investigation. There are also key concerns relating to due process, admissibility of evidence, jurisdiction and interpretation.[56] One consequence of the use of this tool by law enforcement is an inversion of involvement. This is because the DNA databases they use are predominantly made up of samples from family historians and so the data mainly relates to 'white' individuals, 'a racial disparity that is the opposite of disparities in traditional forensic databases'.[57] Several commentators have noted the irony of this in an American society in which ethnicity is a key factor in being involved in criminal investigations.

Fictional Cold Cases

DNA profiling has had an enormous influence on the way that genetic science is conceived of in the popular imaginary. The motif of the 'fingerprint' reinforces the understanding that DNA is unique to each individual and a marker of identification. It makes the invisible visible, with DNA being characterised and to a certain extent anthropomorphised. Conversely, popular versions of DNA profiling have proliferated, and themselves sustain notions of reliability and ease of use.[58] In contemporary crime fiction of all kinds the 'risk' of DNA contamination at crime scenes, and to the criminal, is a given. Clean suits, attempts at scrubbing scenes, and sudden investigative breakthroughs due to scientific evidence, are all standard tropes. In particular the influence of the American show *Crime Scene Investigation* (CBS, 2000-2015) and what has been called the concomitant '*CSI* effect', has raised public expectations of forensic science and especially of DNA evidence. DNA data here is used as evidence in constructing a narrative about an event that happened in the past. In some ways, then, this is the most straightforward use of genetic information in an 'historical' mode. DNA information is brought in to point towards truth and to uncover or reveal wrongdoing. Once again, as with so many examples we have seen, DNA is simultaneous, living in the present investigation as a part of the past event.[59] It is highly imagined, easily troped, and widely recognisable. DNA as evidence here has a relatively direct function, acting as either proof or disproof.

This evidence function becomes more complicated, and more historical, when we consider cold cases. Cold cases have been around for some time as a popular concept. The most famous cold cases include some of the most famous crimes never solved, including the murders of Mary Ann Nichols, Annie Chapman, Elizabeth Stride, Catherine Eddowes and Mary Jane Kelly in 1888 (known as the 'Jack the Ripper' case). The idea of redressing horrors from the past, being able to retrospectively understand the past, is foundational to detective fiction. As Ian Rankin's character Rebus discovers, 'As a detective, he lived in people's pasts: crimes committed before he arrived on the scene, witnesses' memories ransacked. He had become a historian, and the role had bled into his personal life. Ghosts, bad dreams, echoes'.[60] However cold cases as a procedural 'type' have become higher profile, and very much more part of popular culture, over the past two decades. This is due in part due to the increased sophistication of forensic technology. They depend to a certain extent on the sense of a technological, post-genomic modernity, which enables the investigator to understand the past in ways that the original protagonists do not. Evidence was there – and often it was collected – but it could not be interpreted or understood in such a way as to discern the truth.

The importance of cold cases to contemporary imagining of DNA, and in particular its function in reinterpreting and reconfiguring the past is demonstrated by the huge increase in TV shows dedicated to such investigations. *CSI* had a spin-off series *Cold Case* (CBS, 2003-10) which formalised the investigation of past crimes with modern techniques. There is particular interest in the reinvestigation of old cases due to new forensic evidence, seen by the television fictions *Cold Squad* (Canada TV, 1998-2005), *Waking the Dead* (BBC, 2000-2011), *New Tricks* (BBC, 2003-15), *Zettai Reido* (Japan, Fuji TV, 2010-11), *Unforgotten* (ITV, 2015-), *DNA* (Denmark, Nordisk 2020), and true crime documentary series including

Cold Case Files (A&E, 1999-2006; Netflix, 2017), *Cold Justice* (TNT, 2013-15; Oxygen 2017-), *Solved* (Investigation Discovery, 2008-10) and *The Genetic Detective* (ABC, 2020-). The stirring final speech in *The Pembrokeshire Murders* (ITV, 2021) underlines the sense that genetic information can transform our understanding of the past: 'Today's verdicts are a warning to all individuals … Whatever they touch will bear witness against them. Evidence lasts. It does not forget and it cannot lie. And it will always be found by those with the will to find it'.[61] Data is lying hidden, waiting to be found and to change the contemporary world. As discussed above, genetic genealogy has now been used to investigate cold cases, with some success and some ethical controversy. Using genetic databases to match DNA taken from crime scenes as a way to open up new leads is a way of expanding the scope of investigation. It involves genetic family history directly with police procedure, rendering 'historic' DNA valuable. *The Genetic Detective* presents this new approach, featuring genealogist CeCe Moore investigating unsolved crimes using genetic genealogy techniques. In this show the cold case genre has become linked directly to family history and genetic investigation.

The surge of interest in forensic technologies and cold cases can also be seen in crime fiction novels. In 1990 Patricia Cornwell's highly influential novel *Postmortem* introduced an investigative protagonist, Dr. Kay Scarpetta, who is a forensic pathologist rather than a detective. *Postmortem* worked on the principle that forensic investigation could solve a crime without the traditional methods of detection. The novel also introduction DNA profiling, combining innovative detection principles with new genetic technologies. DNA is now a standard investigative technique, enabling the detective finer grained understanding of crime scenes and ensuring that criminals have to be increasingly careful. Val McDermid, one of the bestselling crime writers of the past decade, has an entire series of cold case novels featuring Karen Pirie (2003-). Pirie eventually becomes head of the Historic Cases Unit for Police Scotland, recognising how the investigation of cold cases has become institutionalised in policing practice. McDermid's Pirie novel *Out of Bounds* (2016) demonstrates thoughtfully the impact of DNA evidence on policing, as well as dramatising the problems that genetic information can cause for families. *Out of Bounds* twins two interlinked narratives about DNA: one brings new information for a cold case, the other depends on the new knowledge cheap consumer genetic tests can enable. One of the victims is murdered because he is about to discover, via a commercial test, damaging information about his immediate family. Both crimes are solved through the intercession of DNA information. The murderer is caught by remnants of DNA found on a jacket which causes one of the characters to reflect that 'You'd think these days that everybody would be forensically aware' (p. 479). *Out of Bounds* demonstrates quite how ingrained DNA profiling has become to the imagination of police work, but also how it enables the solving of cases that have been closed for decades. The marketing blurb for *Out of Bounds* uses the genetic form to describe the chaos of the crime: 'finding the answer should be straightforward, but it's as twisted as the

DNA helix itself' (back cover). The novel is self-consciously using DNA in its structure (doubled storyline, the doubling of characters, switched parents) in similar ways to those texts discussed in Chapter 5. This allows reflection upon the complexity of genetic structure whilst also discussing and deploying genetic knowledge as part of the narrative. Similar to the Arnaldur Indriðason novel *Jar City* (discussed in the Introduction) the novel investigates the consequences of expanded genetic knowledge. As with *Jar City*, McDermid's novel reflects the anxieties of contemporary society, and the crimes she outlines are contingent upon their historical particularity. Pirie's understanding of cold cases has a familiar note of melancholia, as she recognises that historic crimes might have been solved had the technology existed: 'The evidence that had nailed her killer simply hadn't been available to them'.[62]

The upsurge of interest in cold case investigation post-2000 coincides with the swift expansion and development of forensic investigation, and also with the completion of the Human Genome Project. These shows suggest that we have the technology to understand violent crime now, and to understand more clearly the events of the past. These series and novels between them contribute to the hardening of a popular truth about DNA, that is, its function as a tool of detection, revealing and opening up to the light long-buried secrets, things that simply could not have been known in the past. These series enable a pre- and post- genetic state of modernity to be understood, insofar as contemporary science interprets the body in different ways to former investigation. The DNA works here as a historical telescope, providing a continuum between then and now, 'living' then and now, having consequence in the past and in the present (and its matter in the present changing the understanding of the past).

Privacy and the Collection of Data

There is an ethical dimension to the means of collecting evidence, revolving around access, legitimation and storage. Any work using human tissue or human remains is subject to this ethical governance, itself overseen by national and international legal frameworks. Indeed the collection of and circulation of DNA evidence is overseen by more law than many other types of evidence, and hence the information has been considered within multiple nexuses of ethics, implication, and legality long before it is interpreted. Previous to the explosion in commercial testing, nearly all DNA evidence had been collected within an ethical framework of some kind. Any scientific collection is subject to ethical oversight, and commercial collection is to a certain extent governed by privacy rules of varying kinds.[63] There are multiple different contexts for this. Each different scientific discipline and locale have different ethical and legal rules for collection.[64] Yet we could argue that, broadly, any collection and interpretation of human genetic material has been overseen by some kind of ethical guidelines. These guidelines have interpreted or implemented the instructions of organisations such as UNESCO (International Bioethics Committee), the European Union, or national governments. Key examples of such governing bodies would be the Nuffield Council on Bioethics (UK, 1991-), the Presidential Commission for the Study of Bioethical issues (USA, 2009-), or the German Ethics Council (2008-). It is only very unusually the case that work appears outside such ethical boundaries – for instance, in the case of He Jiankui who claimed to have used CRISPR technology to 'edit' human twins – and such work is often highly criticised as a consequence for ignoring the international guidelines.[65] Therefore most genetic information generated has an 'unseen' ethical dimension, insofar as it has been developed within an ethical governance framework (although see discussion of Neanderthal DNA, below).

Amendments to legislation regarding ancestry DNA tests were rejected in October 2019 by the National Assembly in France. Indeed, the Assembly took further steps to actively discourage recreational testing: 'L'Assemblée nationale a même été plus loin en durcissant la loi. Un amendement spécifique a été adopté pour interdire formellement la publicité pour les tests génétiques "récréatifs"'.[66] This is part of a longstanding resistance to widespread DNA testing in France. The law is less clear in Germany, but 2009 legislation relating to DNA tests seems to discourage Direct To Consumer DNA testing. Several DTC companies work there, but Ancestry (the largest in the world) is yet to expand into that territory. Each country around the world has different rules and so the companies must adapt their practice. Legally the work is already happening and expanding so fast that governments struggle to keep up.[67]

Governance of matters relating to DNA is slow in responding to swift changes in the market.[68] This is hampered by the number of companies that offer some form of 'leisure' genetic testing expanding so quickly. Around 300 companies currently offer some form of genetic test, and increasingly their product is becoming complex (that is, they offer both ancestry *and* health, or some form of combination). Companies offer tests that cover health, diet, ancestry in various forms, paternity, sports ability, 'talent' (aptitude and intelligence). Some companies will test for infidelity, encouraging the collection of samples without the knowledge of the subject. There are those that offer information about compatibility with potential partners. Many of these companies work in one territory, that is, the USA or the UK, for instance. However their services often can be accessed globally. These testing companies contribute to the wider sense that one might 'know' oneself more completely through understanding one's DNA (see Chapter 6). In this mix are concepts of biological precision, revelation, and a developing sense that our genetic selves are accessible and understandable. Through knowing the microcosmic aspects of the individual we might navigate the world better. The nexus of the new human is a complex interaction of inheritance and newness. The companies participate in an evolution of what the human might be, and how it might be known. Our health, ancestry, abilities, familial connections and taste are all comprehensible, and all interconnected. The historical, ancestry-ethnicity aspect of DNA personal data becomes imbricated in a 'modern' understanding of the genetic self.

Historical ancestry and contemporary health are overlain in the chain of evidence about the individual that DNA provides.

The use of genetic genealogy to solve cold cases is one of the most challenging and strange aspects of the expansion of the genealogical DNA databases. It was totally unforeseen, and this in itself opens up more questions about the collection and storage of DNA information in this way. The speed of expansion of the DNA databases and what can be inferred from them means that legislation is slow to catch up, leading to possible unethical use of the information. Additionally, the transformation of knowledge over such a short period of time means that the implications of the expansion of the databases are only now being understood. The use of genetic genealogy begins to warp the way that DNA might be considered 'evidence'. The imbrication of legal discourses with scientific and heritage concepts means that genetic genealogy becomes part of a movement toward a kind of truth, an uncovering leading to a resolution. It is important to distinguish between the moral ethical aspect here – the discussion of the use of data – and a wider sense of an ethical approach to narrating and understanding the past using genetic means. The use of genetic material as evidence, or for evidence collection, in order to investigate events and crimes from the past brings into focus the historical ethics of DNA.[69] The co-option of genetic genealogy for law enforcement shows how an historical practice that uses and generates genetic information as evidence must still conceive of its own ethical position. Genetic information is demonstrated here to have multiple possible meanings as evidence, and to be interpretable in diverse contexts.

The primary ethical issue here relates to privacy and the ways in which new genetic knowledge might involve individuals in investigations that are nothing to do with them. Millions of users who did not conceive that their 'data' might be manipulated by other organisations had to think carefully about privacy and the implications of the information that they produced about themselves and held on others. They were forced to see their activity and practice as something with an ethical dimension, as having an effect on themselves and others. The second ethical issue involves the usage of information to tell an alternative story to that which the evidence was originally designed for. Historical data is not inert, and the organisation and arrangement of information in particular contexts or for particular purposes – such as criminal investigation – again demonstrates this clearly. Misusing genetic information to investigate crime reveals the ideological template of each new interpretation. A further ethical issue relates to the usage and storage of information. The ethical archive does not exist and hence again the user is forced to recognise that the usage, storage, and manipulation of information relating to the past is an ideological activity.

Furthermore in the contexts just described, genetic history makes us recognise the dead and our relationship and duty to them.[70] Primarily this encounter is through our own genetic link with those who came before us, but also with those who might be given justice due to the intervention of our own DNA. In the examples of cold cases, historical investigation, and the identifying of remains, we find a direct link with the 'other' to whom we have ethical responsibility. DNA hence connects the self in the now with the billions in the past and enables a way of understanding that 'telling' that past is always ethically compromised. However, *not* to tell that past is much more problematic, so the point is to negotiate the responsibility we have to the dead.

De-Extinction and the Ethics of Modernity

On 15 March 2013 the TED educational media organisation, in partnership with the National Geographic Society, live-streamed an event entitled 'Revive and Restore'.[71] 11000 people around the world watched as participants debated and discussed the possibility and the advisability of resurrection biology, also known as de-extinction. Beth Shapiro, a participant in the event, recalled that 'those of us who supported de-extinction [...] hoped that de-extinction would become a tool that the conservation community could add to their existing arsenal of defense mechanisms against contemporary extinctions'.[72] De-extinction challenges historians to engage with and understand genetic science's vision of the world and way of conceptualising biological life as something within time. Geneticists here become restitutive, revisionist historians, curating a way of engaging with the past and reinscribing the past in the present. As well as introducing new objects of study the concept reconfigures the practice of history, as ways of engaging with and understanding the past shift. Evidently de-extinction offers the contemporary world a seeming direct way to engage with creatures from the past via the intercession of genetic science. However, this is hardly unproblematic, this section continues our discussion of ethics by considering the implications for the historical imaginary of the *possibility* of such genetic time-travel.

De-Extinction

The move towards de-extinction has been driven by developments in genomic science over the past 20 years.[73] The genetic work makes it possible to reintroduce cloned versions of creatures that once existed but now do not. There are several types of de-extinction proposed: selective back-breeding, cloning, and genomic engineering. Most commonly discussed over the past decade, driven by technological advance in genomic biology, is genetic engineering:

> by using ancient DNA from museum specimens, material that was previously thought to be unusable, scientists can now sequence the extinct genomes and 'edit' the DNA of closely related species to come up with a genetic blueprint very similar to the extinct forms.[74]

Revive & Restore, a conservation network and global research leader, terms this area the 'emerging field of biotechnology-based genetic rescue', which 'expands ecological restoration capabilities'.[75] There are several lines of work on varying creatures from the Tasmanian Tiger to the Woolly Mammoth.[76] A team in Spain was successful in cloning a Spanish ibex that died due to a lung defect.[77] Many of these creatures were eradicated at the hands of humans (the Tasmanian Tiger was hunted into extinction in the 1930s), so the suggestion is that to reintroduce such species would be restitution as well as possibly useful for conservation. De-extinction proponents argue that such interventions will 'restore lost ecological functions and enhance the diversity of ecosystems'.[78]

Current research on archaic humans such as Neanderthals or Denivosans demonstrates the moral and ethical edges of contemporary DNA work.[79] There has been some theoretical discussion of cloning Neanderthals, although there is no serious work in this field.[80] However new research using ancient DNA in contemporary settings also invokes a model of reanimation. Such work includes the development, through using CRISPR technology and human stem cells, of a technology to grow Neanderthal brain organoids.[81] Organoids are simple three-dimensional cell constructions made in vitro; combining Neanderthal genes into human pluripotent stem cells enables the creation of something akin to archaic human brain tissue. Considering the differences between this and human brain development may allow an understanding of human and possibly primate brain evolution. This type of work enables an understanding of human brain development but raises serious ethical concerns about the nature of the material that is being created, how it is being used to comprehend human evolutionary processes, and the complex power relations which are at play when creating tissue from a long-dead species.[82] Creating tissue from Neanderthal cells in order to find cures for human illness is morally complicated. Treating archaic humans as a repository for profitable genetic information might be considered biocolonialism and certainly has echoes of the historic usage of genetic material from Indigenous peoples (see Chapter 4). Neanderthal de-extinction, even if only certain organs for this type of medical analysis, may not be too far away.[83]

As can be seen from these examples, work in the area of de-extinction raises a number of ethical and (evidently) logistical issues. Quite apart from the question of whether it is advisable (and possible), the re-introduction of species with no social grouping would very possibly lead to extensive problems experienced by the new creatures. We may damage existing populations and introduce disease or other unforeseen problems.[84] Critics suggest that de-extinction takes up resource that might be directed at living better now, and that the moral case (humans should try to reverse what they have done) is not as important as preventing continuing extinction now. Original habitat would also be decayed and changed substantially.[85] This latter is due mainly to human intervention, industrialisation, and climate change. Such a situation demonstrates the historical paradox of living in the Anthropocene. Human intervention has changed the earth, imposing a particular temporal existence on even those creatures that have not existed for centuries. The earth has changed since their extinction events. Further criticisms regarding de-extinction come from postcolonial critics, who argue that the representational consequence of such discourse is to render a binary between modernity and a primitivist, nativist, simpler past, a division that is itself destructive for contemporary ideas of indigeneity. The morality of working on organoids has been widely discussed but introducing a new species into the practice brings in a new set of ethical concerns about consent, human identity, power, and morality.[86] Cloning Neanderthals would raise questions about their rights as types of human. Using Neanderthal DNA in the laboratory highlights the power relations involved in using tissue samples obtained from what are often burial sites.

At the same time many working in the area argue that de-extinction has vital and urgent contribution to make in the context of the climate disaster. George Church, head of a team at Harvard working on de-extinction of the Woolly Mammoth, claimed in a widely publicised quote that the process would help fight global warming: 'They keep the tundra from thawing by punching through snow and allowing cold air to come in. In the summer they knock down trees and help the grass grow'.[87] De-extinction often presents a human-centred ethics which claims to redress the shame of our participation in the original eradication of the species. This is a way of using our new genetic knowledge and expertise to redress historical wrongs, to make amends and atone for the misguided actions of our ancestors.[88] De-extinction thinking would not have been practically possible 20 years ago. As a way of approaching conservation its innovation is to challenge us to wonder 'what if?' It forces the human to consider its intervention into the ecology of the planet and also highlights a type of historicised genetic modernity. We are now sufficiently advanced via genetics to begin to investigate how to recreate and resurrect. Genetics becomes a way of inscribing the past in the present, or, rather, of time travelling.[89] The geneticist becomes someone physically materialising something from the past into our contemporary moment. Even the debate around de-extinction invites us to imagine this apparition.

The impact of the climate emergency on various types on cultural thinking has led to new ways of conceiving of the human and the past. Scholars are increasingly interested in *long duree*, transnationalism, new ontologies and the disruption of 'modernity' as experienced through time and chronology. Wai Chee Dimock cites Ferdinand Braudel in her discussion of 'slow-moving history', arguing that a consideration of 'deep time' opens up a radically new way of thinking about art, culture, and identity.[90] 'Deep time' challenges us to think about how questions of periodisation, time, nationhood, and ontology relate to the nature of humanness and our framework for thinking about 'history'. More radically, Dipesh Chakrabarty has called for a move towards 'species thinking' in the light of the overwhelming evidence of climate change. Chakrabarty's pessimistic point is that humanity has become a 'force of nature in the geological sense'.[91] The advent of the Anthropocene has collapsed the division between 'natural' and 'human' history: 'the wall between human and natural history has been breached' (p. 221). Indeed the increased acceptance of the Anthropocene as terminology and time-period has led many humanities scholars to consider the value of criticism in the light of the climate disaster. Where Paul Gilroy called for the 'new humanism' as a consequence of genetic science, the impact of climate change on humanities thinking has been a call for a move to 'species thinking'. 'Species thinking', like 'deep time', forces humanities scholars to consider 'transtemporality' and to think about 'nonhuman contexts [...][and] suprahistorical temporal frames'.[92] In this context, de-extinction suggests a way of thinking about DNA as anti-temporal, somehow out of time. The practice seeks to collapse human time by intervening through genetic manipulation. In this imagination DNA is somehow contextless, simply being moved from one location in chronology to another. Whilst debates rage about the ethics of transferring animals into a world where their habitat has disappeared, the central idea that DNA is anti-temporal is not discussed. DNA abides, outside temporality – or, at least, human temporality. The possibilities of rewilding through the resurrection of species suggests a way of genetically en-gineering a rejection of human temporality, of challenging our destructive modernity. The reintroduction of species will allow the 'return' of something prehistorical. Something that has been erased from history is now brought back into temporality. This reintroduction of something that had hitherto been absent, through genetic intervention, remakes the world. It suggests a kind of nostalgia that is bound up with contemporary guilt and shame for the current ecological disaster.

Jurassic Park

As a case-study to expand these ideas I am now going to look at the *Jurassic Park* sequence of films (1997-2019). These films dramatise debates about de-extinction and the ethics of genetic engineering. The films work through ideas about the past as commodity, and reflect a continuing concern with the impact of DNA tech-nology on the modern world. The dinosaurs that are created collapse modernity and challenge human definitions of temporality. They are from prehistory but appear in the contemporary, and their very existence suggests the strange and challenging temporality of DNA.

Michael Crichton's 1990 novel *Jurassic Park* rehearses the ethical and technical challenges of de-extinction.[93] In the novel geneticists extract DNA from blood preserved in amber and clone a number of dinosaurs. The DNA is not complete so the clones are not perfect, gaps in their DNA are filled by that of frogs. Hence the creatures are really newly created, and indeed they can utilise some of the genetic characteristics of frogs (they can change sex, which means that they can procreate). The main concern of the project is (initially) education but quickly the financial and combat possibilities of the new creatures and technologies become uppermost. The newly cloned creatures are released into a safari park on an island, with the view that people will be made better through the contemplation of the awesome beauty of the creatures. Soon, though, the dinosaurs are rampaging around de-stroying the park and chasing the humans as prey.

The first film of the novel was highly successful and there is subsequently a long franchise of films from 1993-2020 (six in total, with a CGI-animated series planned). The films were made in two periods, 1993-2001 and then 2015-2020, reflecting two high points of genetic innovation and public interest in ancient

DNA and de-extinction. The films work through ideas about the past as commodity, and reflect a concern with the impact of DNA technology on the modern world. The first three films debate the ethics of cloning and of zoos, and explore the possibilities of working with ancient DNA to create new life. The more recent *Jurassic World* series reflect shifts in DNA technology, concentrating on newer anxieties about genetic engineering, rewilding, and the hybridisation of species (Figure 4.1).

The films suggest that DNA manipulation and the ability to reach into the past will always eventually become monetised. This is the key critique of the first set of films, that the profit motive blinds those running the project to the ethics and possible consequences of their actions: 'You stood on the shoulders of geniuses to accomplish something as fast as you could'.[94] Ian Malcolm accuses the park creators of intentional blindness: 'before you even knew what you had you patented it and packaged it and slapped it on a plastic lunchbox, and now you wanna sell it'.[95] These comments reflect anxieties about the patenting of genes but also suggest that genetic resurrection of the past into the present will always become subservient to profit. The past is commodity, there to be manipulated and profited from. In some ways the message of the films is that the past is just another territory

FIGURE 4.1 Jurassic Park Entrance Arch at the Universal Studios Islands of Adventure

Public Domain, https://commons.wikimedia.org/wiki/File:Jurassic_Park_Entrance_Arch_at_the_Universal_Islands_of_Adventure.JPG.

Credit: Malpass93.

for capitalism to claim, once it has ravaged the contemporary world it will reach back into the past and begin to retrospectively privatise it: 'An extinct animal brought back to life has no rights. It exists because we made it. We patented it. We own it'.[96] In the second set of films the concern is that manipulating the past will enable warfare in the present. Ancient DNA is made combat ready, 'Every bone and muscle designed for hunting and killing'.[97] The agency of the dead, here, is their capacity for violence in modernity. They have nothing other than this in their genetic make up, and that is their perceived value. The communication with the past that genetic science allows is here in the cause of profit, the materials of history used to generate revenue.

Evidently the films themselves participate in the work of turning the DNA of the past into commodity and performance. They themselves work as 'theme parks' of the past in the present, turning profit from the work of the 'dinosaurs'. Protagonist Ian Malcolm captures the movement of the films in the first sequel, *Jurassic Park: Lost World* (1997), itself a happily 'meta' text: 'Oh yeah, 'Ooh Ahh' that's how it always starts. But then there's running and screaming'.[98] They make the work of presenting the past into a theme park, something for entertainment rather than serious consideration. The geneticist in charge of the work, Henry Wu, reminds the owners of the park in *Jurassic World*, 'you didn't ask for reality; you asked for more teeth!' In the same way, the film technologies have worked to amplify the creatures and to grow their threat.

In the original three movies the concerns of gene editing surrounded the ethics of doing something simply because it was possible. The lesson of the trilogy was that to mess with nature was to invite disaster. At the conclusion of the first sequence of films the creatures have taken over the islands that they were bred and shown on, aggressively rewilding them. In these locations only the ruins of theme park buildings represent human society, echoes and gaps of capitalist attempts at owning creation. *Jurassic Park* ends with the sentiment that 'life will find a way', suggesting that the creatures will evolve into their environment.[99] Modernity will be circumvented on these two islands, as the contemporary ecology meets the prehistoric, and human history becomes irrelevant. This is reflected in the im-placability of the creatures in the face of the human. Humanity, time, modernity and capitalism are irrelevant to dinosaurs that live according to different temporalities.

The most recent tranche of films explores splicing of characteristics but also more complex issues such as animal socialisation and evolution in different ha-bitats. The protagonist Owen Grady has pioneered communicating with a group of cloned Velociraptors, with the dominant female 'Blue' showing throughout two films a clear sense of engagement and even community. In contrast, the 'genetically modified hybrid' *Indominus rex* is raised in captivity and isolation, and as a consequence is presented by the film as entirely irrational and dysfunctional.[100] Communicating with the creatures, rather than othering them as savage, is the means to a particular type of progress. Despite this the dinosaurs that are created throughout – including a final *Indoraptor* hybrid in *Jurassic World: Fallen Kingdom*

(2018) – collapse modernity and challenge human definitions of temporality. They are from prehistory but appear in the contemporary, and their very existence suggests the strange and challenging temporality of DNA. The creatures are sublime, jarring, and beyond human time. Every film has a moment in which the humans are taken aback by the majestic beauty of the creatures, but these scenes also suggest an awe at something that should not exist *now* ('First time you see them, it's like a miracle').[101] This is articulated in a short exchange in *Jurassic World*. When Vic Hoskins, the park security claims 'Extinct animals have no rights' Grady replies cynically: 'They're not extinct anymore, Hoskins'.[102] The legal definition of de-extinct creatures is something that has been debated since the idea was first introduced.[103] Hoskins argues that these creatures do not have identity, they are somehow new and non-things. Grady points out that the impossible now exists, and must be conceptualised as something with rights, legal definitions, and agency. Through the intervention of genetics human time is challenged and warped, as what was then is now.

This intervention is transformative and cannot be changed. Ian Malcolm, the character who connects the first and second sequence of films, claims at the conclusion of *Jurassic Park: Fallen World* that a kind of rewilding has taken place as a consequence of de-extinction, and that this 'change' is part of the process of genetic manipulation: 'Genetic power has now been unleashed and of course, that's going to be catastrophic'.[104] He points out that 'These creatures were here before us. And if we're not careful, they're going to be here after', concluding: 'We're going to have to adjust to new threats that we can't imagine. We've entered a new era, welcome to Jurassic World'. For Malcolm the ecological makeup of the world has shifted – in this case, due to innovations in genetic technology. Malcolm points to the fears surrounding 'genetic power' and the 'catastrophic' consequences of using this. De-extinction and the manipulation of DNA has led to a fundamental shift in the way that the world is configured, and this is the warning of the films. The final comments also reflect contemporary awareness of the climate disaster and the Anthropocene. Human behaviour has changed the planet, and it may be that the species can no longer survive in the face of 'new threats' that we have brought upon ourselves. Human time will end, and other species will reassert their dominance. Even the notion of an 'era' becomes a failed attempt at owning and understanding this new moment. Human attempts to understand the past in terms of sections or sequences diminish the enormity and non-human aspect of what is happening now. Human history and historying is over. Through blind action the species has invoked new horrors, 'fundamental' change that it has no way of redressing or reversing. The dinosaurs represent an implacable self-created future apocalypse that cannot be ignored or prevented.

Yet the idea of 'Jurassic World' also mobilises the entire film sequence's concern with non-Metropolitan and marginal spaces, and Malcolm's concern that the primitive and the premodern will invade the contemporary. The location of each park is on the edges of centralised, civilised world. It is a place of savagery and wildness, primitive and paradisical. The tourist-dynamic of the parks works on the fact that the visitors are visiting the past as a landscape as well as a safari of creatures. The use of military-style helicopters and vehicles throughout the films, particularly by those 'rescuing' the stranded tourists, underlines the type of intervention that is being made into the savage space. The jungle and its darkness emphasise the similarity to Vietnam in the cinematic imagination, with an implacable foe striking violently from the darkness. Each film involves the failure of modern technology, from radios to electric fences, precipitating the (often literal) invasion of the savage and animalistic into the space of the present. This is evident in several films in which characters return to ruined spaces of former park buildings, rewilded by the jungle and increasingly hostile to moderns. The rewilding suggests that without human interference the world will repair itself or return to an Edenic state of 'innocence' (read as an unsophisticated, premodern existence).

The *Jurassic Park* sequence of films both reflect contemporary anxieties and also suggest a matrix of responses to the debates around de-extinction. A renewed creature is born into a world that has changed, they are out of time, anti-historical and entirely new. Like the creature in *Frankenstein*, the new animals will have no way of being socialised or educated. They will have no 'past', appearing without context or parent and alienated, alone as a consequence. Genetic science here allows the development of a new/old species, renewed and reconfigured according to human wishes. The process effects a kind of posthuman challenge to the centrality of human experience. Symbolically the 'new' dinosaur *Indoraptor* is killed by the raptor 'Blue' at the end of *Jurassic World: Fallen Kingdom* by being impaled on the horn of a *Ceratopsian* fossil. In this sequence three distinct types of 'dinosaur' are seen, a clone of a 'real' creature, an engineered 'new' creature, and a 'real' fossil. The levels of development are clear here, as the 'new' creature is vulnerable to the wiles of the clone and the brute force of the 'real' dinosaur. However the *Ceratopsian* is in a museum, to be seen, owned, and understood; the other creatures are alive and unbiddable, violent and terrifying.

Notes

1 Anton Froeyman, *History, Ethics, and the Recognition of the Other* (London and New York: Routledge, 2019), p. 5. This would be different in tone and purpose from existing bioethical work on genomics.
2 See the discussion of the variety of ethical historical positions in Marnie Hughes-Warrington, *Big and Little Histories: Sizing up Ethics in Historiography* (London and New York: Routledge, 2021).
3 There have been some, relatively brief, attempts at articulating an ethics here but from the scientific, rather than historiographic, perspective, see Lori B. Andrews et al., 'Constructing Ethical Guidelines for Biohistory', *Science* 304: 5668 (2004), 215–6.
4 Archive Fever (Chicago, IL: University of Chicago Press, 2017), p. 1.
5 Thomas Laqueur, *The Work of the Dead* (Princeton, NJ: Princeton University Press, 2015).
6 Crandall and Martin, 'The Bioarchaeology of Postmortem agency', *Cambridge Archaeological Journal* 24:3 (2014), 429–35.
7 Thomas J. Parsons et al., 'Large Scale DNA identification: The ICMP experience', *Forensic Science International* 38 (2019), 236–44.

8 Outlined for instance in Sarah Wagner, *To Know Where He Lies: DNA Technology and the Search for Srebenica's Missing* (Berkeley, CA: University of California Press, 2008).

9 See the essays in *Human Remains and identification: Mass violence, genocide, and the 'forensic turn'*, ed. Elisabeth Anstett and Jean-Marc Dreyfus (Manchester: Manchester University Press, 2015).

10 See for instance Jay D. Aronson, 'Humanitarian DNA Identification in Post-Apartheid South Africa' in Wailoo, Nelson and Lee, *Genetics and the Unsettled Past*, pp. 295–314.

11 Francisco Ferrandiz. "The Ethnography of Exhumations", in Francisco Ferrandiz, ed. Necropolitics: Mass Graves and Exhumations in the Age of Human Rights. Philadelphia: University of Pennsylvania Press, 2015.

12 'Identifying Democracy: Citizenship, DNA, and Identity in Postdictatorship Argentina', *Science, Technology, & Human Values* 41:6 (2016), 1037–62 (p. 1038).

13 S. Cardoso et al., 'Contribution of forensic genetics to the recovery of historic memory of the Spanish Civil War', *Forensic Science International* 1:1 (2008), 454–56; Francisco Ferrándiz, 'The return of Civil War ghosts: the ethnography of exhumation in contemporary Spain', *Anthropology Today* 22:3 (2006), 7–12 and Layla Renshaw, *Exhuming Loss: Memory, Materiality and Mass Graves of the Spanish Civil War* (California: West Coast Press, 2011).

14 'The missing, the martyred and the disappeared: Global networks, technical intensification and the end of human rights genetics', *Social Studies of Science*, 47:3 (2017), 398–416 (p. 398).

15 Esther Lee et al., 'MtDNA origins of an enslaved labor force from the 18th century Schuyler Flatts Burial Ground in colonial Albany, NY: Africans, Native Americans, and Malagasy?' *Journal of Archaeological Science* 36 (2009), 2805–10.

16 Mai-Linh K. Hong, 'Get Your Asphalt Off My Ancestors!: Reclaiming Richmond's African Burial Ground', *Law, Culture and the Humanities* 13:1. (2017), 81–103.

17 Her institute is the Restitution Study Group, http://rsgincorp.com/home [accessed 14 May 2020].

18 Jennifer A. Hamilton, 'The Case of the Genetic Ancestor' in Wailoo, Nelson and Lee, *Genetics and the Unsettled Past*, pp. 266–79 (p. 271).

19 Quoted in Eric Bock, "Social Life' of DNA has unique power', *NIH Record* 71:4 (2019), https://nihrecord.nih.gov/2019/02/22/social-life-dna-has-unique-power [accessed 14 May 2020].

20 *In Search of Our Roots: How 19 Extraordinary African Americans Reclaimed their Past* (New York: Random House, 2009), p. 10.

21 Alondra Nelson, 'The social life of DNA: racial reconciliation and institutional morality after the genome', *British Journal of Sociology* 69:3 (2018), 522–37.

22 'DNA, reconciliation and social empowerment', *British Journal of Sociology* 69:3 (2018), 546–51.

23 Kim TallBear, 'Genomic articulations of indigeneity', *Social Studies of Science* 43:4 (2013), 509–533. See also See Amade M'Charek, *The Human Genome Diversity Project* (Cambridge: Cambridge University Press, 2005) and Darryl Leroux, 'We've been here for 2,000 years': White settlers, Native American DNA and the phenomenon of indigenization', *Social Studies of Science* 48:1 (2018), 80–100.

24 On the rise of 'racist science' in relation to DNA and methods of resistance, see Adam Rutherford, *How to Argue with a Racist* (London: Weidenfeld and Nicolson, 2020).

25 Alondra Nelson, *The Social Life of DNA* ().

26 Michael Lynch, Simon A. Cole, Ruth McNally, Kathleen Jordan, *Truth Machine: The Contentious History of DNA Fingerprinting* (Chicago, IL: University of Chicago Press, 2010).

27 A.J. Jeffreys, V. Wilson and S.L. Thein, 'Individual-specific 'fingerprints' of human DNA', *Nature* 316 (1985), 76–9.

28 Mark A. Jobling and Peter Gill, 'Encoded evidence: DNA in forensic analysis', *Nature Reviews Genetics* 5 (2004), 739–51 (p. 739).

29 Robin Williams and Paul Johnson, *Genetic Policing: The Use of DNA in Criminal Investigations* (London and New York: Routledge, 2008), p. 97.

30 'Eureka moment that led to the discovery of DNA fingerprinting', *The Observer* 24 May 2009, https://www.theguardian.com/science/2009/may/24/dna-fingerprinting-alec-jeffreys [accessed 22 June 2020].

31 See Colin Me Haverson, 'What Results Should Be Returned from Opportunistic Screening in Translational Research?', *Journal of Personal Medicine* 10:1, 13 (2020), doi: 10.3390/jpm10010013. Thanks to Jay Clayton for this reference.

32 Sheila Jasanoff, 'The Eye of Everyman: witnessing DNA in the Simpson trial', *Social Studies of Science* 28:5–6 (1998), 713–40.

33 Rhonda Wheate, 'The importance of DNA evidence to juries in criminal trials', *The International Journal of Evidence and Proof* 14:2 (2010), 129–45 Lisa L. Smith, Ray Bull and Robyn Holliday, 'Understanding juror perceptions of forensic evidence', *Journal of Forensic Sciences* 56:2 (2011), 409–14.

34 Information from 'National DNA Database statistics', https://www.gov.uk/ government/statistics/national-dna-database-statistics [accessed 22 June 2020].

35 Helen Wallace, 'The UK National DNA Database', *Science and Society* 7 (2006), S26–30.

36 Caleb Hutton, 'Partial skull bone found last year belonged to Gold Bar man', *HeraldNet*, 10 August 2019, https://www.heraldnet.com/news/palm-sized-piece-of-skull-identified-as-that-of-gold-bar-man/ [accessed 15 August 2019].

37 Sarah Zhang, 'She was found strangled in a well, and now she has a name', *The Atlantic* 29 July 2019, https://www.theatlantic.com/science/archive/2019/07/belle-well-dna/594976/ [accessed 15 August 2019].

38 Tina Hesman Saey, 'What FamilyTreeDNA sharing genetic data with the police mean for you', *ScienceNews*, 6 February 2019, https://www.sciencenews.org/article/family-tree-dna-sharing-genetic-data-police-privacy [accessed 18 September 2019].

39 Caleb Hutton, 'Man guilty of 1987 murders solved with genetic genealogy', *The Herald*, 29 June 2019, https://www.heraldnet.com/news/man-guilty-of-1987-murders-solved-with-genetic-genealogy/ [accessed 19 August 2019].

40 https://snapshot.parabon-nanolabs.com/ [accessed 19 August 2019].

41 https://snapshot.parabon-nanolabs.com/ [accessed 19 August 2019].

42 Sara Gilcore, 'A local company is helping catch cold case criminals', *Washington Business Journal*, 18 October 2019, https://www.bizjournals.com/washington/news/ 2019/10/18/a-local-company-is-helping-catch-cold-case.html [accessed 21 October 2019].

43 Natalie Ram and Jessica L. Roberts 'Forensic genealogy and the power of defaults', *Nature Biotechnology* 37 (2019), 707–8.

44 Yaniv Erlich, Tal Shor, Shai Carmi, 'Identity inference of genomic data using long-range familial searches', *Science* 362 (2018), 690–4 (p. 690).

45 Erlich, Shor, Carmi, p. 690.

46 Erlich, Shor, Carmi, p. 690. See Sarah Zhang, 'Most people of European Ancestry can be identified from a relative's DNA', *The Atlantic*, 11 October 2018, https://www. theatlantic.com/science/archive/2018/10/golden-state-killer-genealogy/572545/ [Accessed 15 August 2019].

47 Ram and Roberts, 'Forensic genealogy', p. 708.

48 Christine Guest, 'DNA and Law Enforcement: how the use of open source DNA databases violates privacy rights', *American University Law Review*, 68:3 (2019), 1015-52 (p. 1016).

49 J.W. Hazel, E.W. Clayton, B.A. Malin and C. Slobogin, 'Is it time for a universal genetic forensic database?', *Science* 362: 6417 (2018), 898–900.

50 Benjamin E. Berkman, Wynter K. Miller, Christine Grady, 'Is it Ethical to Use Genealogy Data to Solve Crimes?', *Annals of Internal Medicine*, 169:5 (2018), 333–4.

51 Christi J. Guerrini et al., 'Who's on third? Regulation of third-party genetic interpretation services', *Genetics in Medicine*, 12 August 2019, https://doi.org/10.1038/s41436-019-0627-6.

52 Natalie Ram and Jessica L. Roberts 'Forensic genealogy and the power of defaults', *Nature Biotechnology* 37 (2019), 707–8.

53 M.J. Radin, *Boilerplate: The Fine Print, Vanishing Rights, and the Rule of Law* (Princeton, NJ: Princeton University Press, 2013), pp. 11–12.

54 Nathan Scudder et al., 'Massively parallel sequencing and the emergence of forensic genomics: Defining the policy and legal issues for law enforcement', *Science & Justice* 58:2 (2018), DOI: https://doi.org/10.1016/j.scijus.2017.10.001

55 Natalie Ram, 'Genetic Privacy After Carpenter, *Virginia Law Review* (forthcoming 2019) http://dx.doi.org/10.2139/ssrn.3265827.

56 Nathan Scudder et al., 'Policy and regulatory implications of the new frontier of forensic genomics: direct-to-consumer genetic data and genealogy records', *Current Issues in Criminal Justice* 31: 2 (2019), DOI: https://doi.org/10.1080/10345329.2018.1560588.

57 Erlich, Shor, Carmi, 'Identity inference', p. 690.

58 Barbara L. Ley, Natalie Jankowski and Paul R. Brewer, 'Investigating *CSI*: Portrayals of DNA testing on a forensic crime show and their potential effects', *Public Understanding of Science* 21:2 (2012), 51–67.

59 See Jay Clayton, 'Genome Time', in *Time and the Literary* ed. Karen Newman, Jay Clayton, and Marianne Hirsch (New York: Routledge, 2002), pp. 31–59.

60 Ian Rankin, *Black and Blue* (London: Orion, 1997), p. 365.

61 *The Pembrokeshire Murders*, ITV, January 2021, Episode 3.

62 Val McDermid, *Out of Bounds* (London: Sphere, 2017), p. 474.

63 Stuart Hogarth and Paula Saukko, 'A market in the making: the past, present and future of direct-to-consumer genomics', *New Genetics and Society*, 36:3 (2017), 197–208.

64 José Roberto Goldim, 'Genetics and ethics: a possible and necessary dialogue', *Journal of Community Genetics* 6:3 (2015), 193–6,

65 Dennis Normile, 'CRISPR bombshell: Chinese researcher claims to have created gene-edited twins', *Science* 26 November 2018, https://www.sciencemag.org/news/2018/11/crispr-bombshell-chinese-researcher-claims-have-created-gene-edited-twins [accessed 18 March 2020].

66 Guillaume de Mordant, 'L'Assemblée nationale rejette les tests ADN généalogiques et durcit la loi', *La Revue française de Généalogie*, 4 October 2019, https://www.rfgenealogie.com/s-informer/infos/nouveautes/l-assemblee-nationale-rejette-les-tests-adn-genealogiques-et-durcit-la-loi [accessed 7 October 2019].

67 See discussion in Stuart Hogarth, Gail Javitt, and David Melzer, 'The Current Landscape for Direct–to Consumer Genetic Testing: Legal, Ethical, and Policy Issues', *Annual Review of Genomics and Human Genetics* 9 (2008), 161–82 and Stuart Hogarth and Paula Saukko, 'A market in the making: the past, present and future of direct–to–consumer genomics', *New Genetics and Society* 36:3 (2017), 197–208.

68 Andelka M. Phillips, 'Only a click away — DTC genetics for ancestry, health, love… and more: A view of the business and regulatory landscape', *Applied and Translational Genomics* 8 (2016), 16–22.

69 Onora O'Neill, *Justice across Boundaries* (Cambridge; CUP, 2016).

70 See Edith Wyschogrod, *An Ethics of Remembering* () and Elizabeth Grosz, *The Incorporeal* (New York, NY: Colombia UP, 2016).

71 https://www.ted.com/tedx/events/7650 [accessed 30 July 2019].

72 Beth Shapiro, *How to Clone a Mammoth: the Science of De-Extinction* (Princeton, NJ: Princeton UP, 2015), p. 169.

73 Beth Shapiro, 'Pathways to de-extinction: how close can we get to resurrection of an extinct species?', *Functional Ecology* 31 (2017), 996–1002.

74 Ben Minteer, *The Fall of the Wild: Extinction, De-Extinction, and the Ethics of Conservation* (New York, NY: Columbia University Press, 2018).

75 https://reviverestore.org/what-we-do/ [accessed 30 July 2019].

76 Charles Y. Feigin et al., 'Genome of the Tasmanian tiger provides insight into the evolution and demography of an extinct marsupial carnivore, *Nature Ecology & Evolution* 2 (2018), 182–92.

77 Jose Folch et al., 'First birth of an animal from an extinct subspecies (*Capra pyrenaica pyrenaica*) by cloning', *Theriogenology* 71 (2009), 1026–34.

78 Ben Minteer, 'Is it right to reverse extinction?', *Nature* 509 (2014), 261.

79 Sariah Cottrell, Jamie L. Jensen and Steven L. Peck, 'Resuscitation and resurrection: the ethics of cloning cheetahs, mammoths, and Neanderthals', *Life Sciences, Society and Policy* 10:3 (2014),https://doi.org/10.1186/2195-7819-10-3.

80 Zach Zorich, 'Should we clone Neanderthals? The scientific, legal, and ethical obstacles' *Archaeology* 63:2 (2010), 34–41.

81 Ariana Remel, 'Neanderthal-like 'mini brains' created in lab with CRISPR', *Nature* 11 February 2021, 376–7. I owe this point to Marnie Hughes-Warrington.

82 Julian J. Koplin and Julian Savulescu, 'Moral Limits of Brain Organoid Research', *Journal of Law, Medicine & Ethics* 47:4 (2021), 760–7.

83 Rebecca Wragg Sykes, *Kindred* (London: Bloomsbury, 2020).

84 Shlomo Cohen, 'The Ethics of De-Extinction', *Nanoethics* 8 (2014), 165–78.

85 Michael J.L. Peers et al., 'De-extinction potential under climate change: Extensive mismatch between historic and future habitat suitability for three candidate birds', *Biological Conservation* 197 (2016), 164–70.

86 Insoo Hyun, J.C. Scharf-Deering and Jeantine E. Lunshof, 'Ethical issues related to brain organoid research', *Brain Research* 1732 (2020), https://doi.org/10.1016/j.brainres.2020.146653.

87 Hannah Devlin, 'Woolly mammoth on verge of resurrection, scientists reveal', *The Guardian* 16 February 2017, https://www.theguardian.com/science/2017/feb/16/woolly-mammoth-resurrection-scientists [accessed 7 August 2019].

88 Stephanie S. Turner, 'Open-ended stories: Extinction narratives in Genome time', *Literature and Medicine* 26:1 (2007), 55–82.

89 Jay Clayton explores this via his concept of 'genome time' in 'Time considered as a helix of infinite possibilities', *Medical Humanities* 47 (2021), 185–92.

90 *Through Other Continents: American Literature across Deep Time* (Princeton, NJ: Princeton University Press, 2008), p. 5.

91 'The Climate of History: Four Theses', *Critical Inquiry* 35:2 (2009), 197–222 (207).

92 Ben Morgan, 'After the Arctic Sublime', *New Literary History* 47:1 (2016), 1–26 (4).

93 Shapiro, *How to Clone a Mammoth*, p. 56.

94 *Jurassic Park* (Steven Spielberg, 1993).

95 *Jurassic Park* (Steven Spielberg, 1993).

96 *Jurassic Park: Lost World*.

97 *Jurassic World: Fallen Kingdom*.

98 *Jurassic Park: Lost World* (Steven Spielberg, 1997).

99 *Jurassic Park*.

100 *Jurassic World* (Colin Trevorrow, 2015).

101 *Jurassic World: Fallen Kingdom* (J.A. Bayona, 2018).

102 *Jurassic World*.

103 Norman Wagner et al., 'De-extinction, nomenclature, and the law', *Science* 356: 6342 (2017), 1016–17.

104 *Jurassic World: Fallen Kingdom*.

5

IMAGINATION

Genetic science has long been engaged with by artists, writers, filmmakers, poets, and musicians.[1] Similarly, popular culture is incredibly important in shaping wider understanding of genetic science.[2] In the past two decades or so writers such as Zadie Smith (*White Teeth*), Kazuo Ishiguro (*Never Let Me Go*), Octavia Butler (*Xenogenesis* series), Amitav Ghosh (*The Calcutta Chromosone*), and Margaret Atwood (*Oryx and Crake*) have all published influential work pondering the ways that genetics presents a challenge to humanness.[3] Films and television series from *X-Men* (film series 2000-) to *Rampage* (Brad Peyton, 2018) and *Orphan Black* (Space/ BBC America, 2013-17) have explored identity, anthropocentrism, and genetic modernity. For these writers and film-makers genetic science allows the exploration and articulation of transgressive and politically enfranchising identity. Yet these texts also demonstrate the anxieties attendant upon being 'postgenomic', exploring the edges of genetically-defined humanness in order to reflect upon the possibilities but also the problems of this seeming new state of being. Critics such as Josie Gill, Lara Choksey, Jay Clayton, Jackie Stacey, and Clare Hanson have persuasively argued for the imbrication of genetic knowledge with aesthetic representation. As Gill outlines, 'contemporary fiction explores, actively participates in, and might inform, the biofictional formation of racial ideas in genetic science'.[4] Choksey asks 'how can narrative redirect flows of interest and information towards imagining a different set of correspondences between biology and society', seeing in engagement with DNA the potential for new configurations.[5] This chapter participates in this critical engagement with genetic aesthetics, outlining some of the ways that artists have interrogated genetics over the past two decades and how this relates to imagination of the past. Some are fascinated with the representational possibilities of genetics, and in particular the ways in which DNA might enable a new type of modernist, anti-bourgeois, queer 'realism' to be developed. In the cases of Marc Quinn, Bernadine Evaristo, and Ali Smith, engagement with genetics enables this rethinking of representational tropes. This itself entails a reckoning with historical

DOI: 10.4324/9781003052975-6

practices and aesthetics that seeks to situate the 'now'. Poets and rappers use genetics to articulate new political configurations, particularly in relation to discourses of race, and consider genetics as a means for interrogating form and identity in relation to the past. For all these writers, 'postgenomic' is a template to be rejected, a means to attempt to control and subject which should be critiqued. Finally, I look at ways in which genetics has been imagined as a means for accessing a past fraught with anxiety, a mode of time-travel that might disrupt the present.

Materialising DNA

This section considers the ways in which writers have investigated the physicality of genetics as a means for challenging normative assumptions about identity, history, and realism. In particular, the section looks at representational tropes are put under pressure by meditations upon genetics, and in particular how discourses of the real can be challenged through using and engaging with genetic science, leading to potential reconfiguration of the self and society. The experimental poet Christian Bök has for some years been working on the 'Xenotext', a project to encode poetry into DNA strands. The principle is to inject a created strand of DNA (itself the first 'part' of the poem) into an organism which itself will then incorporate the 'poem' into itself and then 'write' a protein in response: 'I'm genetically engineering a bacterium that won't just archive my own text in its DNA, but also becomes a machine for writing a poem in response'.[6] In this work Bök has attempted to materialise the relationship between genetic proteins and poetry.[7] His work is suggestive for the ways in which art and genetics might interrogate one another. This section, then, outlines ways in which artists have reflected upon the new possibilities of genetic science to reconfigure profoundly the form and material of their art.

Genetic Realism: Marc Quinn

Marc Quinn's *A Genomic Portrait: Sir John Sulston* (2001, National Portrait Gallery) raises many of the key considerations regarding memory, identity, genetic realism and the past that a postgenomic aesthetics would need to grapple with. Sir John Sulston was famous at the time the portrait was made for his public involvement in the Human Genome Project, in particular asserting the importance of it being a free, public project. He would later share the Nobel Prize for Physiology of Medicine in 2002. The image was produced in collaboration, a plate of Sulston's DNA propagated in agar jelly and mounted on stainless steel. Commissioned by the National Portrait Gallery (NPG) as part of a partnership with the Wellcome Trust, the image was intended to provoke debate about the nature of portraiture, of art, and of science. Displayed in March 2001, the piece was a collaboration between the Wellcome Trust and the NPG.[8] The work was a way of developing a partnership relating to science communication. It also augmented the NPG's commission budget (which is fixed yearly) and enabled them to invest in what they saw as an important movement, that is, Young British Art. The piece foregrounds a kind of genetic realism, that is, a mode of representation that seeks to as accurately as possible represent the original. Ken Arnold wrote in *Tate* magazine that 'Quinn's portrait insistently asserts the need to think much more laterally ... [and] fuses the ideas of portraiture and biography'.[9] The viewer looks at Sulston's DNA, sees themselves reflected in the stainless steel, and 'reflects' upon the matter of life, of representation, and of art.

The image is highly unusual and challenges the institution in many ways regarding its conservation. It is 'alive' in a way that nothing else in the NPG is. This raises ethical and legal issues. The NPG wrote to Sulston in 2006 seeking his consent for storage and use of his tissue, after the passage of the *Human Tissue Act* in 2004.[10] When the image was being planned some members of the development committee worried that the work would have to be registered as a toxic hazard.[11] The material aspect of the image has proven challenging to conservation. Its own journey through the past years has not been totally frictionless. The image's life suggests something interesting about the physicality of the object but also the ways in which biological material might be 'fixed'. This story tells us a lot about how 'art' and science might combine and interrelate, and some of the curatorial, archival, and practical issues associated with such engagement. The work 'means' differently now through its temporal and material ageing, it has achieved a kind of 'maturity' of meaning.

Firstly, there is the question of what the image actually is made of. In 2004 Helen White's conservation overview reports that 'In November 2003 when this work was examined for loan to Sheffield it was noted that the appearance of the work had changed. Extra shapes were beginning to form in several places within the agar layer'.[12] It is clear that the agar layer, the central focus of the image, was deteriorating: 'By this time the original had deteriorated quite dramatically; the crystals had grown and the layer thinned to around 1/6 of its original thickness and consequently darkened'.[13] This was because the agar is drying out. Quinn, Sulston, Kathleen Soriano and White looked at the work, and decided that it needed attention. In particular, the decision was made to ask the Sanger Institute to provide another plate of DNA that would be sealed. An email from Kathleen Soriano mentions 'The work has changed ... Marc's assistant, Angela, has developed a new way of sealing it'.[14] Hence the technical aspects of the artwork have subtly changed, and demanded a new process, and a small but material shift of the work to become something else. They use a resin 'chosen by Quinn because it did not interfere either physically or visually with the bacteria', however

> In order to avoid the original deteriorating further (though it was already unexhibitable), resin was poured onto that also. It immediately reacted very badly and during the period of hardening the sides and bottom of the perti [sic] dish distorted, the agar darkened further and was pulled away from the dish, bubbles and cracks formed.[15]

So the *original* – which is already something that cannot be displayed – essentially disintegrates, quite impressively. One of the four 'spares' (five plates were provided at commission) is used to replace this. The 'new' image becomes classified as

'6591.1', and the spares 'named sample 1, 2 and 3 in the drawer, but also 6591.2, 6591.3, 6591.4'.[16] Quinn's image is apart from other things a riff on replicability, and the materiality of the 'human' in the portrait. Already – three years after its first showing – it is itself undergoing a process of decay, changing in relation to temporal movement, becoming something new – and needing to be replaced, remade. Given, too, that the picture raises issues of 'originality' in art, the fact that the *original* has been replaced – albeit by a plate that was provided for just this occasion – raises a question about the nature of the piece. Is the 'new' piece the same as the old? The DNA is the same, the materials similar, but the age of the objects and their configuration is a new one.

The resin coating applied in 2004 appears to 'fix' the agar well enough. That said, subsequent examinations of the 'new' image reveal lines and 'tide marks', splits and crystallisation effects. White's report from August 2009 argues that this 'suggests the agar is drying out … the one currently labelled Sample 2 has crystals forming in it'.[17] In 2011 a circular line was reported across the 'new' image's agar, along with various small dots; by 2012 this had disappeared, although the spares were continuing to show signs of decay.[18] White concludes 'Why these have deteriorated more than the framed version is not know' and warns 'they may be an indication of what could happen to the framed version in the future' (p. 1). In 2016 when White looked at the picture, she reports 'The prime object remains in good and displayable condition. However, it is probably more fragile than it was'.[19] However, she records that the samples 'continue to display serious defects' (p. 1), adding 'It continues to be a mystery as to why the spare plates have deteriorated more than the framed version' (p. 1).

This had been anticipated to a certain extent. Minutes of meetings relating to the image and related exhibition demonstrate that the discussion of the sustainable nature of the image had been ongoing from before it was executed.[20] However it is not clear how this was resolved, and the ensuing discussions in 2004 suggest that the issue was undecided to a certain extent. Soriano's emails to Ken Arnold of the Wellcome Trust in 2004 outline ongoing discussions about the upkeep of the plates and hence the sustainability of the image. Arnold and Soriano had been involved in the original commission, and were clearly aware of potential problems with the piece in the future: "there was always the element of the unknown when we began on this project and I think that it was always something of a tall order expecting Marc not to be interested in the element of change, let alone our limited knowledge of the medium!"[21] Sulston himself emails Soriano and Arnold in May 2004 that he had hoped to 'look after things personally for a while' but his situation has changed; however, 'Of course we would eventually have reached this point anyway, so it's not bad to be confronting it now'.[22] During the commissioning process Sulston had repeatedly discussed the limited 'lifespan' of the plate. In his email Sulston outlines the technical challenges: 'The plating procedure entails a number of steps that are routine for the subcloning group, but require both skill and appropriate reagents for success. To safeguard the object in perpetuity will need continued access to those skills and reagents'.[23] The different technical demands of the lab and the conservation centre are illustrated here. The demands of preserving this piece are complex and technical. Unlike many similar concerns within data preservation about interoperability and hardware, for instance, this relates specifically to the material nature of the piece itself. Whilst this is not uncommon in relation to conceptual art, the wider contexts of science communication and the NPG's commitment to preserving work for the nation make the discussion pointed.

These 2004 emails were obviously prompted by the re-examination of the piece and the now relatively pressing need to develop a plan for the future of the artwork. Soriano urges Arnold, and the Wellcome Trust, to commit to holding 'John's DNA clone library for us in perpetuity' that 'would also retain the Wellcome's link with the piece'.[24] In 2004 Arnold reported to Soriano that the Sanger Institute had agreed 'to 'keep alive' Marc Quinn's portrait of John - i.e. periodically refreshing the work with more of John's 'clone library'.[25] This is further underlined by Bronwyn Terrill from the Sanger Institute in 2010, and emails from Soriano to Arnold suggest that Quinn was very clear that the Sanger had agreed to maintain the work 'as it was recognised at the point of commission that this work would deteriorate over 5-10 years'.[26] Email exchanges between the Sanger Institute and the NPG during 2010 and 2012 outline in detail what kind of work is needed on the plates and the agreement between the institutions, which seems never to have been put on a formal footing.[27] Yet in 2013, Fay Blanchard, evidently trying to piece together what the situation was, emailed to Arnold 'It seems the departments involved were then restructured and the trail went cold [...] The Sanger now don't feel they can support the ongoing replacement of the plates'.[28] The Sanger, according to Blanchard, no longer have a stock of Sulston's processed DNA; as she points out 'If this is no longer the case them [sic] we have a sizeable problem/ need to re-think the piece entirely'.[29]

How far does any of this matter, in terms of the piece itself and the materiality of the work? Much conceptual art is bedevilled with issues of conservation and even challenges the nature of such sustainability. A great deal of conceptual art from the late 1990s, too, is interested in ephemerality and the inbuilt decay of materials. Indeed, Quinn's most famous work, *Self* (1991 and ongoing), a sculpture of his head using his blood fixed in frozen silicon, reflects upon the conditions of its own preservation. Yet this image has a counter-discussion at its heart, a reflection upon something seemingly fixed, almost transhistorical or anti-temporal. Quinn characterised it as 'a portrait of his parents and every ancestor he ever had, back to the beginning of Life in the universe'.[30] This sense of deep time seeks to give the artwork a status as a fixed temporal point. DNA gains the status of something accurate, authentic, and – to a certain extent – unchanging.

The image is challenging in terms of the ways in which we understand the postgenomic human. It suggests an identity outside of the normative body, genderless and without race, class, past, something non-temporal. The absolute modernity of the piece is part of its original purpose. Writing about the image in the *Guardian*, Jonathan Jones suggested that:

> this is a kind of biological photography. As in a portrait photograph, a specific trace of someone has been fixed permanently. Through the most sophisticated modern means, Quinn's work revives the primitive impulse at the heart of portraiture: preservation of a person.[31]

Technically this is not true – the 'specific' trace has not been fixed permanently, or preserved. The implications for what the material trace of Sulston fixed by the image exactly is, then, are manifold. The image is beyond the means of the gallery to keep and has an in-built expiration date. Something that, when first created, was hailed as admirably modern, technologically 'sophisticated', and, above all, permanent, is revealed by archive to be fragile. The seemingly solid message of modern science melts into air. Quite literally, at the centre of the image of 'Sulston' is a decaying, drying, lifeless set of bacteria, and agar that rather than being continually on show must be attended to and remade repeatedly. It points to a melancholic, rather than heroic, version of the scientific portrait, something fading rather than vanquishing time. The material problems experienced by the image prompt us to once again recognise the historical embeddedness of objects. The precarious quality of the image and its ongoing decay work against the modernity inherent in its original conception. If DNA makes a key contribution to our understanding of ourselves as human, we need to locate that understanding precisely. What we know now is contingent on being now, and the hindsight afterlife of the Sulston portrait reminds us that our understanding of the human is evolving and temporally contingent. DNA may seem to eschew historicity, to exist outside of human historical time, but its understanding, representation, and interpretation is resolutely structured within modes of knowing, memorialising, and archiving. Is the image Sulston, or a 'version' of him? Quinn's portrait raises multiple issues for postgenomic art and aesthetics. Primarily, how might genetic realism be understood as an aesthetic that offers 'exact' and authentic replication? It is seemingly possible to exactly recreate someone. Furthermore, how does form relate to content in postgenomic art, and how does this express a fundamental fragility about representation? Does the 'image' gain some kind of inflection or new dimension through being materially biological? Finally, how might this type of work be conserved and curated into the future? As a side issue, the ownership and responsibility of the image also provides a challenge to 'normal' ways of understanding, archiving, and owning art.[32]

Experimental Form in Ali Smith's How to Be both

Ali Smith's experimental novel *How to be both* is similarly concerned with using genetics as a means for challenging normative representation modes and seeking a 'new' realism. The novel plays with doubling and the Mobius strip of touching-not touching interrelationships over time.[33] The novel is divided into two narratives, printed in different order randomly, and each beginning 'one'. There is a simultaneity and a temporal ordering of this. Each experience of the book may be different, but only around two poles. The book's format switches back on itself, reorganising itself for different readers. In approaching them the reader imposes some kind of temporal ordering, a hierarchy of precedence which can never be unpicked. This challenge to temporal normality of the novel is common to some contemporary fiction, seen in Sarah Waters's *The Nightwatch* for instance. Like Waters, Smith has a desire to challenge heteronormative structuring and in one sense the dual narrative is simply an experimental strategy to highlight the novel's concerns with control, linearity, and becoming. Hence Smith's work challenges history by articulating its own queering of historical discourse. There is the briefest of jokes about this when two girls have a conversation about revising for fact-based Biology and Latin exams by putting the information to the 'tune of some song they both know': 'They'd been standing in the corridor outside history' (p. 92). The two girls are excluded from the academic subject of History, seeking a way to subvert knowledge by singing it.

The two sections are very distinct, but related (although never touching). One details the grief of a contemporary girl, George, and her attempts to understand the death of her mother. The other, more experimental, section, voices Francesco del Cossa, a fifteenth-century artist that Smith imagines as a girl brought up as a boy, and further imagines as a kind of disembodied ghost in the book, a presence watching the contemporary moment, attempting to understand, and telling the story of their own development and aesthetic work. The novel shifts and plays with duality and temporality. In one memory, George recalls discussing versatility and migration with her mother, whilst watching a documentary about historical trains on TV 'and because it is an interesting programme, she is simultaneously watching it from the start on catch-up on her laptop' (p. 40). This ability to consume art (and art about the past) in multiple times, the simultaneity of existence and historical event, is something Smith interrogates closely in the text. Art might both extinguish historical difference and render it entirely, playing on its own continuum. It might be forgotten, and its maker marginalised by history, or painted over. George and her mother discuss simultaneity when considering art: 'But *it* isn't, George says. Because that was then. This is now. That's what time *is*' (p. 104). Her mother disagrees: 'Do things that happened not exist, or stop existing, just because we can't see them happening in front of us?' (p. 104). This interrogation of linear ('that was then. This is now') with multiplicity, the suggestion of other ways of being and of seeing, are part of the novel's subtle queering of sight and aesthetics. The title, *How to be both*, is highly suggestive, implicating gender, life and death, young and old, as binaries that might be challenged as easily as George changes her name from Georgia. This fluidity is part of Smith's wider aesthetic practice, but here it is sharpened into a meditation upon history and existence. In one of the many moments when the link between template and lived reality merge, George's mother says 'Its as if that map they gave us is nothing to do with the experience of being here' (p. 62).

This doubling and consideration of simultaneity is clearly influenced by genetic science, and the novel includes several sections of direct engagement with DNA. Whilst the majority of the book is in prose, the more experimental section uses a kind of poetry at the beginning and end of this narrative (see image). As the voice swims into existence the prose figures itself as a helix. As with the doubling of the narrative, the text's formal innovation turns around DNA. Indeed, the experimental aesthetics of the novel are enabled by this engagement with the double helix. Smith's points about life and art and simultaneity are hence governed by these formal interventions. Her engagement with the contemporary novel form – largely linear, largely in straight prose – queers it and challenges such tropes of normativity, whilst re-rendering a different type of form. Smith posits an alternative, an other, a translation of the traditional format.

The spiral-script fragmented poetry of the account resolves itself into prose, and then splits into poetry at the end of the narrative. The artist is seemingly reborn:

> Ho this is a mighty twisting thing fast as a
>   fish being pulled by its mouth on a hook
>   if a fish could be fished through a
>   6 foot thick wall made of bricks or an
>   arrow if an arrow could fly in a leisurely
>   curl like the coil of a snail or a
> star with a tail if the star was shot
>   upwards past maggots and worms (p. 189).

For some readers this i the first aspect of the text they see. The text curls around itself. The text is defined by the blank page around it, as much given shape by its length and data than its meaning. This 'mighty twisting thing' seems to suggest life, a movement, a journey of some kind – but is also a clear reference to the helix that the prose is describing physically. It is both fast and 'leisurely', impossible and very real. In the final section of the text the prose disintegrates again and there are three pages of helix curl, ending with:

>   the curve of the eyebone
>   of the not yet born
>   hello all the new bones
>   hello all the old
>   hello all the everything
> to be
>   made and
>   unmade
>   both (p. 372)

The text unravels, falls apart, seems to lose its structure – although it similarly gains another set of structures, the helix and the internal logic and prosody of verse.

Like the ellipsis at the end of Roberts's poem, discussed below, the novel has no ending. It opens up to the blankness of the page and the reader's mind. Bear in mind, though, that for half the readership this lack of an ending simply moves to the next narrative, where for half it is the conclusion of the novel. All the readers live in this potentiality of 'both' before they open the novel, like a Schrodinger's cat somehow alive and not alive. The body of the novel is itself reimagined, presented as a fragmented thing, incomplete but full of potentiality.

Smith chooses to dwell on ambiguity and strangeness, spectrality and in-effability. The novel transforms into verse, a change which is physical (the shift of the words and margins) and conceptual (the shift from the driving narrative of prose to the fragmentation of poetry). It is simultaneously a novel and a poem, the flexibility of language defined by its formatting and relationship to that around it (the blank space of the page). It is put into tension with the non-language of the blank page, made to mean in comparison to that which does not mean. This final section challenges time ('the not yet born'), conceiving of 'everything' being a state of flux 'made and/ unmade/ both'). This dislocation suggests that the dou-bling is inextricably constant, the moment is 'made/ and unmade' simultaneously and continuously. The decision to turn 'made and/ unmade/ both' into three lines suggests a revolving around the words

> made and
> > unmade
> > > both

The eye gets to 'and' and then has to return back on itself to 'unmade', then again back on itself slightly to 'both'. It isn't necessarily a helix but it is a revolution, a spiral of kinds. This section seems to dramatize Latour's thoughts about simultaneity, quoted below. The voice is old and new, then and now, whole and fragmented, 'both'. Smith's conclusion seems to dramatise the potentiality of understanding history as genetic, as not linear and progressive and purposeful but winding and strange and simultaneous. The modernity of the novel, with its concerns of digital identity, online shaming, contemporary aesthetics and the like, are in tension and dialogue with older aspects of art, memory, form. If we conceive of our relationship with the 'past' as something febrile and continuous, a 'shout to the sky' which is both physical and unmade, biological and imaginative, then we reconfigure the way that we live. The novel's epistemology argues that we do not 'know' but we do un-derstand, so we need to embrace this complexity and oddness.

DNA testing and Form: Bernadine Evaristo

That aesthetic engagement with consumer DNA testing is well advanced is de-monstrated by Bernadine Evaristo's *Girl, Woman, Other*, the book that shared the Man Booker Prize for 2019.[34] The book is overtly and explicitly about race and racial identity in the United Kingdom. *Girl, Woman, Other* concerns twelve loosely connected women, subdivided into chapters focusing on groups of three that are more clearly related, either in friendship or by family. What brings the women together is Amma's play *The Last Amazon of Dahomey*, due to be performed at the National Theatre. On the one hand this represents the storming of the institutions, as Amma's radical lesbian vision is allowed onto the stage of the National (and as Evaristo herself becomes the first Black British woman to win the Man Booker prize). On the other it presents a type of assimilation, a recognition of an inclusive and diverse modern country.

Each woman has some non-white heritage, and the short sections outline their lives and identities and the ways in which they have been formed by their experiences as non-normative people in the United Kingdom. In short, clipped style Evaristo explores the legacies of immigration on the bodies and the lives of these women. Some challenge their place and some leave, but all are formed – apart from one, the white Penelope who has great privilege and overtly racist attitudes. Evaristo's project is made explicit in the opening chapter in which radical playwright Amma, her 'woke' daughter Yazz, and her great friend Dominique discuss lesbian separatism, intersectionality, institutional racism and various other crucial contemporary political contexts. That chapter sets the framework for the novel's exploration of the politics of identity and black experience. Evaristo's central point is that the United Kingdom is ethnically diverse, families are complicated, and that these stories should be heard. Her chapters present alternative familial structures, reaching back into the past but without a structured linearity or heteronormative ordering.

At the conclusion of the novel Penelope, who discovers early in her life that she was adopted, is prompted to take a DNA test by her daughter. The name Penelope is that of Odysseus's wife who waited for him and the symbolism here is that the character herself comes home to her family, her belonging, after a lifetime of being lost 'all her life' (p. 446). Penelope's test allows Evaristo to aestheticize the experience of seeking for roots but also the revelation that it might enable. Planning to 'surprise' her husband, 'it didn't quite turn out as expected' (p. 446). Rather than a normative, standard outcome she discovers extensive African heritage in her profile. The results mean 'Penelope's suffering from post-traumatic stress disorder' and her world shifts: 'this was the science that was the deepest, most secret part of herself, and there was a collision between who she thought she might be and who she apparently was' (p. 446). Evaristo here articulates the disruption of self that the DNA tests enable, the invasion of the present by genetic data from the past and a shaking of the contemporary person. Whilst 'post traumatic shock' is popularly used as a way of discussing responses to revelation it also suggests that the new information has a traumatic and dislocating effect on this character. There is a difference between aspiration ('who she thought she might be') and 'who she apparently was', and this is established by science that reaches to the 'deepest, most secret part of herself'. This aspect of her is hidden even to herself. Whilst looking for her original family, she is instead confronted by a new genetic identity. The trauma, the 'shock', is that 'the test didn't provide answers, it confronted her with questions' (p. 447).

Racist Penelope must reconfigure her selfhood and understand the ways that her whiteness has been unmarked, although her responses are circumscribed by her attitudes – she thinks her ancestors must have been 'nomads roaming over the continent killing each other before the British demarcated regions into proper countries and thereby imposed discipline and control' (p. 448).[35] She wonders whether she should 'become one of those Rastafarians and sell drugs' (p. 448). However, when she meets her birth mother this feeling melts away because 'who cares about her colour? […] in this moment she's feeling something so pure and primal it's overwhelming' (p. 452). Indeed, the meeting between the two destroys culture and society and renders them 'primal', their DNA and genetic match so strong that 'it's like the years are swiftly regressing until the lifetimes between them no longer exist' (p. 452). Temporality is confounded in this concluding moment of the novel, and experience is created anew: 'this is not about feeling something or about speaking words/ this is about/ being together' (p. 452). Much as the prose of Smith's novel reflected both a structure and a flux, Evaristo's engagement with genetic history imagines a combining and a fragmentation, a reconfiguring into something new. The complexity of the body's temporal construction is key here, something that Latour pointed out; we are never modern and never archaic, but something else entirely. As Rita Felski argues, 'Cross-temporal networks mess up the tidiness of our periodizing schemes, forcing us to acknowledge affinity and proximity alongside difference, to grapple with the coevalness and connectedness of past and present' (Context Stinks! p. 579). DNA has the potential to disrupt and reorder, both aesthetically and physically.

Poetry, Hip Hop, and Postgenomic Imaginaries

This section considers the ways that poets and rappers and artists have responded to the biomedical gaze, and the ways in which they have dissented or built upon this seemingly new way of conceiving of the human and its relation to the past. They consider what it means to live in a postgenomic way, and the possible implications for writing, art, and living. On the one hand, being postgenomic seems to offer a kind of connectivity, a way of understanding the *human* within time and our relation to the past in new and innovative fashion. On the other, poets increasingly see postgenomics as a way of attempting to control the body, to impose and close down meaning. They celebrate fragmentation and strangeness, disintegration and metaphor, and highlight throughout that language – whether it be genetic data, or mathematics, allusion or motif – is a slippery, strange thing to be celebrated and mistrusted.

May Swenson's 1968 poem 'The DNA Molecule' anticipates the duality and non-linear motif that contemporary artists respond to. Figuring DNA as Duchamp's *Nude descending a staircase* Swenson suggests 'She is descending and at the same/ time ascending and she moves around herself […] She is a double helix mounting and dismounting' (ll. 5-6, 10).[36] Swenson suggests that science has just caught up with Duchamp's representative practice and his ability to conceive of something that is mobile and multiple, several times and one at the same time. Similarly several of the modern poets' use of DNA and intertextuality interrogates notions of straightforward lineage and inheritance. Rather than think about linearity they open up spaces of echo and uncanny haunting, using DNA as a motif for a complicated relationship with the past. However, they also all express a particular modernity, a post-genomic sense that things have changed ('*Now*, as your mistress strips for bed/ her body is already mapped' writes Roberts, my emphasis). They have access to particular knowledge that has been unlocked by a technological shift in the past decades. Their particularity, their historical specificity, is keenly felt. DNA is connective and links the individual to multiple others in the past. The poets interrogate the discourses of history whilst queering them through this bodily connection: 'a history that reckons in the most expansive way possible with how people exist in time, with what it feels like to be a body in time, or in multiple times, or out of time [...] a *queer* history'.[37] The way that DNA infracts upon history in these poems and tracks demonstrates this point and emphasizes Dinshaw's understanding of the queerness of temporal existences.

Why has poetry and rap been so receptive to, and creative in, thinking about postgenomic identity? Poets and rappers are conscious of the interrelation between form and content, possibly more than other writers. Certainly, the tension between line and poem, between the technicality of language and its meaning, is highlighted in verse and rap. Imagining DNA as metaphor and thing, something multiple and concrete, material and ephemeral, gives poets a way in to thinking about the ghostly spectrality of humanness, and how this might be reflected in verse.[38] Poets and rappers are interested in roots and migration, in the ways in which identity might be rendered historically. Each poet or rapper uses their form to interrogate the seeming fixity of DNA and to present, instead, a kind of alternative. To contest this, writers rest their practice in conceiving of the human as both historically contingent and specific, but also somehow attached to a species or a family that is older and stranger. They celebrate a transmission between then and now. Poetry understands this consonance and difference, the resonance of the then in the now. These poems and tracks are modern but archaic, much as the human is understood as a modern subject now but shares form with those from the past.

Ghostly poetics: Sullivan, Duffy, Kunial, and Symons Roberts

Hannah Sullivan's 'The Sandpit After Rain' compares DNA articulation to aesthetic composition. Her conception of the body so constructed allows complex movement between the micro and the macro: 'All cancers were once benign/ Then the DNA forgets its prosody'.[39] This leads to disintegration, 'And cells divide interminably:/ The raddled beauty of doggerel'. DNA here is imagined as having 'prosody', rhythm and sound, moving from controlled (benign) form to the interminable chaos of doggerel – which has a kind of unpleasant 'beauty'. The descent of DNA from 'high' form to aged, populist verse leads to the decay of the body and finally death. It is, however, unending – 'divide interminably' – suggesting a kind of looping chaos once the beauty of the body's normative organisation has been disrupted. Sullivan suggests that biotechnology renders the minutiae of the body beautiful, or, at least subject to aesthetic judgement. The poet uses overlay and equivalence here, not metaphor or simile: 'Stained under a microscope,/ An ovary is Venice at sunset, 'Too beautiful to be painted' said Monet'. Here the ovary *is* Venice (when under a microscope); cancer cells gain the 'raddled beauty of doggerel'. Sullivan updates the standard tropes of using the poem to stand in for the body, blazon verse rendering the human known for centuries, for a medicalised, genetic understanding of the role of poetry in expressing the species. Yet the ovary/Venice overlap suggests the body so investigated as being something sublimely beyond art, despite being captured and framed under medical instruments.

Carol Ann Duffy's poem 'Richard' is similarly concerned with the materiality of the body and how it relates to poetics. Written on the occasion of the re-interring of Richard III's body at Leicester Cathedral, duffy's poem ventriloquizes the king and reminds us of the strangeness of the body rendered within an historical gaze. The poem is shot through with a sense of ghostly, spectral doubleness. Richard's existence and material definition is played around with by the verse, which is loaded with suggestive moments. The poem refuses clarity, instead working in a mode of suggestion and strangeness. Richard's status as genetically defined, then, is key to understanding the verse. They are, like braille, both empty and full of meaning, an indentation and absence that can be understood by touch. Richard can only be engaged with on a physical level, 'My skull [...]/ emptied of history' (ll. 2-3). Duffy echoes herself Seamus Heaney's 'Bog Queen', a poem about the digging up of bodies in the peat in Ireland: 'I lay waiting/ between turf-face and demesne wall [...] My body was braille/ for the creeping influences [...] and I rose from the dark'. The metaphor of braille therefore links the two bodies and they are figured as impressions, outlines.

Key to the poem's working through of ghostliness and spectrality is its form. Duffy writes a sonnet, a poetic form that has been used by poets from the sixteenth-century onward. It is one of the most-used poetic forms in the Anglophone tradition, with famous practitioners including Shakespeare, Wordsworth, Barrett-Browning, Spenser, and scores of other writers. It is a challenge as a form, given its precise form of 14 lines. Sonnets also tend to be about consciousness and the difficulties of expressing selfhood in verse. Writing a sonnet requires skill, but it also *de facto* involves an engagement with 400 years of Anglophone poetic history. It cannot *not* echo the multiple versions that have come before, or seek to meet with and change this tradition. Even those who radically undo sonnets, like Elizabeth Bishop, still find themselves working within the framework of the poetic template. This is an historical form, then, understood as part of an ongoing discussion, back and forth, between now and then, about the relationship between language and structure in poetry. A sonnet therefore has a DNA of its own, building blocks that are shared between texts hundreds of years apart. Duffy's use of the sonnet clearly refers to Shakespeare, and his example both of poetry and in 'making' the collective memory of Richard III. She uses the sonnet to conceive of the difference between physical template and life, but reminds us of the open-endedness of structure by introducing ellipsis and hyphens at the end of lines ('loss –'; 'Dead…', ll. 9-10). This extremely weak enjambment allows the line to float away from the body of the poem into the blank white of the page.

The keynote of Duffy's poem is absence/ presence, a meditation upon the interrelation between then and now that DNA itself dramatizes. The voice of Richard conceives of a flexible time: 'or I once dreamed of this, your future breath/ in prayer for me, lost long, forever found;' (ll. 11-12), whilst also conceiving of 'The end of time – an unknown, unfelt loss', l. 9). Contradictory and strange, the poem attempts to understand how to manifest this ghostly presence that is both physical and imagined, how to account for the uncanny oddness of the bones being made to live and having an imagined and a physical connection to the contemporary moment. How does Richard III mean, now? He can only live if he is 'read', if the braille is interpreted. The physical body means very little without our contemporary interpretation, effected through genetics.

Similarly Zaffar Kunial's 'Self Portrait as Bottom' conceives of the way that genetics might render the body measurable. 'Let's get down to numbers', Kunial suggests when describing the results of his DNA test, 'What could be more prosaic?'[40] Rather than the foundational and organisational poetic line suggested by Sullivan's imagining of DNA, Kunial sees the rendering of genetic information in data as something blunt, non-poetic (insofar as it is prose), regular and uninteresting. 'Prosaic' suggests the antithesis of poetry, that this way of portraying the self will lead to a realisation that the human is essentially normal and unmusical. If genomics presents a kind of full-spectrum realism or a way of rendering the human, Kunial's poem suggests that this is aesthetically uninteresting. However, the actual numbers in the poem are highly suggestive, and behind them lie all manner of complex identity issues. The numbers are not neutral, but a mode of interpretation that Kunial seeks to understand.

Kunial's poem struggles to understand the relationship between his DNA and his sense of self. Both Kunial and Sullivan express something about the strange and almost uncanny relationship between data and material body, using metaphors of language to attempt to understand how something imagined can also be something real. Kunial's poem also engages with some of the ideas expressed by Quinn's *Genomic Portrait*, as he looks to understand the 'realism' of his DNA and the actuality of the relationship between this material stuff (literally, his saliva) and his imagined self. 'Self Portrait as Bottom' contributes to a subgenre of poems working through the implications of the auto portrait and issues of representation.[41] Kunial's is one of the first to contemplate the newly defined genomic self, and his poem therefore confronts the contemporary reality of the body in 2018, expressing the modernity of the postgenomic. He perfectly outlines the ambiguities and the anxieties of the modern subject, defined historically through genetics and imagined in relation to data. He is now and then, flexing between ancestry and contemporary. The numbers are a way of expressing this transitive quality, and they are more 'real' as representation of him than the sample of saliva he looks at in the test tube.

Bottom, the weaver character from *A Midsummer Night's Dream*, is given the head of an ass by the fairy Puck in order to mock Hermia, who has been drugged to fall in love with Bottom. Bottom's movement is from the lowest class to the lap of luxury, and the joke in it is that Bottom never knows that he has been changed. *A Midsummer Night's Dream* is a play obsessed with alterations, terrifying mobility, and the possibilities of transformation. The play presents altered states as something to be feared, but Bottom's pompous self-importance means that he lacks the self-knowledge to understand this. The words 'Bottom, thou art translated' are delivered by one of his friends, Peter Quince, when he appears with the head of an ass. The full line generally reads 'Bless thee Bottom, Bless thee. Thou art translated!' The re-petition of 'bless thee' demonstrates Quince's conviction that this is magical, possibly devilish work, and that Bottom needs to be protected from it. His transformation is from one thing into something entirely different, 'translated' meaning carried from one state to another. The joke, further, is that he has an ass-like quality in his character, and that the challenge to anthropocentrism is a way of enabling this to be expressed. He is also, ultimately, the fool – laughed at by the audience, by the fairy king Oberon (and his human equivalent Theseus). The play's jokes with Bottom are about self-knowledge, and the differences between external and internal.

Kunial's poem begins with a version of Bottom's line 'O I am translated'. There are textual variants in Shakespeare's 'original' writing of this line, though, which remind us of the insubstantiality of language in presenting a 'genealogy' of meaning in a text's afterlife. The 'O' comes from a variant version of the text (in which 'translated' becomes 'chang'd'), so Kunial is already pointing out flaws in afterlife, textual allusion, echo and intertextuality. The translation is as much from one mode (drama) to another (poetry); Kunial himself is moved from poetry (humanness) to prose (numbers, ethnicities). Meaning is somewhere between the two aspects, original and variant.

> O I am translated.
> The speech of numbers.
> Here's me in them
> And them in me. (ll. 1-4)

Kunial's self is here imbricated with the information, to the point that when he looks at his saliva in the test tube he looks not just at himself now but also 'past'.

Kunial looks at the test tube with his saliva in and finds it strange, other, odd. The glass reflects his face (he 'saw an elongated/ face of me, staring past/ my drool', ll. 12-14), clearly an updating of the contorted reflection in John Ashbery's famous 1975 poem 'Self-portrait in a convex mirror' (itself a response to Parmigianino's 1524 image of the same name).[42] Kunial's abstracted face, the 'bottom' of the test tube, reflects his expression 'trying to summon [...] the unconnected/ unspeaking dead' (ll. 15, 17-18). Unlike the doomed artist in Ashbery's poem, unable to con-nect, Kunial has a newly modern way of engaging his past and understanding himself – his DNA: 'Me. Or so/ the science and the blurb says' (ll. 19-20). He points out the problem of having his selfhood explained to him by marketers and knowledge gatekeepers. He explains the uncanny strangeness of DNA, something he 'ridiculously [...] looked to see in that U-/ shaped test tube' (l. 11). Kunial expected to see something related to his identity that 'I've felt at some level', 'a kind of abysmal underneathness/ or usness, under the heights/ of language' (ll. 57-9). The link between his parents, their DNA, and him is something unknowable and inexpressible. The link is so deep and not-known that he can barely express it, and recognises that he will not actually get an answer despite the promises of the 'blurb'. Yet there is something there, inherent, linking him with his past; something that he *knows* but cannot understand or see.

Walter Benjamin reminds us of the duality of the text in time, how its multi temporality leads to complexity and multiple meaning: 'A translation issues from the original, not so much from its life as from its afterlife'.[43] For Benjamin, the original and the translation always sit in tension with one another:

> It is evident that no translation, however good it may be, can have any significance as regards the original. Nonetheless, it does stand in the closest relationship to the original by virtue of the original's translatability; in fact, this connection is all the closer since it is no longer of importance to the original.[44]

A translation, or an adaptation, or a response or echo – all have a life as this 'afterlife', running simultaneously but never touching the life of the original. In the same way, Kunial is somehow himself and a version of himself, translated, as Bottom was, into something *else*. Yet which is the echo? Is Kunial the original, or the adaptation?[45] The DNA, as is shown, comes first, historically and chronologically. It links him to a past that he was not part of, that he somehow echoes and reflects. He is the afterlife of this, given illumination by the DNA that he submits to be read. This tension between things provokes a crisis of sorts: 'O'. This particular letter, opening the poem, renders absence and presence simultaneously. Like the DNA it is both shape and letter, sound and text. The letter O works typographically because of the white space that it is not, a constant circling of meaning/non-meaning. The next letter it is not, in this instance, is 'I'. The absence and the self, the interaction between nothing and me, is the crucial crisis of genetics. DNA is everything; DNA is nothing. Kunial's body is both imagined and real, his DNA is there and not there, holding in tension the helix, data and self, now and then, other and portrait. This doubleness and duality, the binary tension, is somehow an aesthetics. As Benjamin writes, each element is in relationship with the other, but simultaneously different. Self and other are in tension forever with the aesthetics of other things, but also changed when a new aspect appears.

Kunial's poem, outside of its title and opening quote (which is not acknowledged as such), uses Bottom's status as a rude, marginalised social figure. He is split between two types – human and animal. He is also non-normative, challenging both social ordering in his relationship with Titania and also anthropocentrism.

Similarly the poet is 'split. *50% Europe./ 50% Asia*'. The line break accentuates the duality which is now institutionalised in his body through the interpretation of science. Kunial, through his use of a DNA test, becomes something new, something else – possibly subverting the ordering systems that had been put in place to control social interrelations. Just as the space of the forest in *A Midsummer Night's Dream* stands for a dangerous possibility and transgression, for Kunial the test tube he spits in might provide freedom, but also involve the untethering of the self from its ordering boundaries. In *A Midsummer Night's Dream*, both the actors and the lovers who use the forest as a place for escape and possibility end up scared of it and fleeing. Such a new space is not for the faint-hearted. To see oneself in this new context, to understand through the writing of 'self' that you might be different than you had imagined, is a challenging psychological thing to do.

Refusing to be simply defined by the 'prosaic' figures and percentages Kunial imagines a mythic background:

> And from my dad?
> *48% Asia South*. Which as good
> as says that my father's
> folk were converts in the near
> past, perhaps lower caste, perhaps
> believers in the many, in sky gods
> cast in Sanskrit, or heavy Buddha,
> or puckish forest figures,
> winged gandharvas. (ll. 35-43)

Kunnial bluntly points out the lack of diversity in the DNA test he has taken. Rather than the European part of his profile, which is broken down to Great Britain, Ireland, Scandinavia, Finland, Italy, the test does not have the ability to present him with fine-grained accounts of his Asian ancestry.[46] In place of any account the poet presents his own versions of what '*Asia South*' might mean, conjuring visions of gods and theologies. The brief and precise use of 'puckish' takes us back from Asia to the world of Shakespeare's play, reminding us that sprites and fairies are not necessarily English, and also suggesting that Kunial's DNA shares Puck's magical ability to 'put a girdle about the earth'.[47] He makes an equivalence between Puck and 'gandharvas', figures from Hinduism and Buddhism. This equivalence, however, is attenuated: 'puck*ish*'. We also recall that in *A Midsummer Night's Dream* the battle between Oberon and Titania is caused by her love for an Indian boy, a changeling child given to Titania by his mother. Does Kunial somehow conceive of himself as the child, and his parents as Oberon and Titania? This seems to be certainly in the shadows, as the poem turns to the geographies of his parents' DNA history 'almost meeting', 'but not quite'. Like Quinn's work, the notion is of a version of the self being presented which is accurate, and at the same time unique, whilst also being somehow generic.

Like Marc Quinn's work, Michael Symons Roberts' poem 'To John Donne' (2004) was created as part of a direct response to John Sulston. The poem opens with an epigraph from Sulston: 'when you patent a gene,/ you are enclosing a part of me, the shared landscape'.[48] Similarly to Quinn, then, Roberts finds in the figure of Sulston a means to focalise discussion of the implications of the post-genomic condition. The verse challenges John Donne and particularly his poem 'Elegy XIX: To His Mistress Going to Bed' (first printed in 1654). This poem, an address about a woman undressing, is a complex work considering power and ownership. Donne is generally conceived of as a very fleshy poet, concerned with the body and its physicality in relation to the divine. Roberts in particular uses the most famous section of the elegy:

> Licence my roving hands, and let them go,
> Before, behind, between, above, below.
> O my America! My new-found-land,
> My kingdom, safeliest when with one man mann'd,
> My Mine of precious stones, My Empire,
> How blest am I in this discovering thee!
> To enter in these bonds, is to be free; ('Elegy XIX', ll. 25-31)

Roberts's updating directly addresses Donne and refers to the body newly under-stood, newly gendered, remade by genomic science: 'Now, as your mistress strips for bed,/ her body is already mapped,/ its ancient names a cracked code' (ll. 1-3). If Donne's female body is a mystery Roberts suggests it has been solved, updated, modernised. Her gender has become immaterial but in the aesthetic of the poem it is still important. As Harraway writes, 'If DNA signifies "life itself" in the semiotic orders of biotechnology, synthetic DNA is especially open to realising the future, and to realizing profit from your investment in that future'.[49] This manifestation of biopower seeks to understand and survey the (female) human, to control and order it. Older, premodern versions of the (male) gaze that constructs, seen in Donne's poem, give way to modern discourses of science that control and commodify.

The body of the lover is known in both poems, understood and read by using governing metaphors and contemporary concerns. In Donne the body is under-stood by his attempts at imposing meaning upon what is up to this point un-known, his desire to explore a need to control and secure the blank space. In Roberts the body is known at a remove. On the one hand, the biomedical gaze would look at the individual woman and 'read' her. On the other, she is simply part of a species that is understood as a mass. Every woman would look similar and can only be understood in comparison to every other woman. Her individuality, lost in Donne, is again erased but this time due to a modern biomedical con-ception of the human. She is both unique and multiple. Donne's misogyny is forgiven in some ways by Roberts because of his insistence in being in the mo-ment, whereas the valuation of the human by gene patenting commodifies us now and forever. Roberts's calling back through the past is an attempt at conceiving of what is particularly *now*. Yet both poets describe the same thing. They simply use different means. Knowledge here is flexible. The object – the objectified – is the lover, she is simply known in different ways in different times. Gender difference here in verse becomes something that can be reinscribed in modern poetry through the intercession of genetic science. Early modern writers 'blazon' or trope the female as land fresh for discovery; contemporary scientists render the human as data. The body is conceived of in different ways. This forces the poet to understand a kind of feedback loop, recognising his own historical embeddedness as he sees the historical specificity of Donne. This is why the poem begins with 'Now'. This is not necessarily progress, but difference. Both express a type of realism, historically rendered and historically specific. The body is never a body, but always a construct, imagined and desired and dreamed by (mainly male) interlocutors.[50]

Roberts's concern is that the 'new found land' that Donne imagined is already 'paced out,/ sized up' (ll. 4-5). He says to Donne 'this atlas of hers is no mystic book', suggesting the replacement of faith with a degrading rationality (l. 11). The biomedical gaze constructs her instead, 'mastered by medics […] a textbook of disease' (ll. 8, 12). She is understood in new ways, her 'charts are held on laptops […] laid bare' (ll. 7, 8). Her nakedness is mathematical and data-driven rather than physical. Roberts seeks a way out of this reification of the flesh. He attempts to short-circuit the biomedical gaze, asking that Donne and his lover move from metaphor to reality: 'Do you care? Does she? What/ can it matter at this fleet May dusk' (ll. 28-29). His request is that the two ignore the biological description and understand the moment: 'Let your hands, and hers, lead us/ in love's mass trespass' (ll. 37-38). In place of the enclosed spaces of the New World Roberts suggests the popular movement against such boundarisation (the Mass Trespass of 1932). Borders should be crossed, and the lovers should be allowed to move towards one another and integrate, 'claim back with whispers/ the co-ordinates of bodies' (ll. 39-40). Finally, the poem disintegrates as the lovers become data:

> the co-ordinates of bodies: TTA,
> GAG, TGT, CCC, ATC, TGT (*this is,*
> *yes, a litany*) CTG, GAG, TTG … (ll. 39-41)

The poem ends with this 'litany' of DNA and then an ellipsis. Rather than itself being end-stopped, concluded, fenced off, part of a template, it opens up. The genome is not to be concluded, bound, but opened and played with.

Another way of reading this conclusion is to suggest that the verse 'becomes' DNA – the biological and the aesthetic switching places. The base acids turn into verse, they engage with prosody and metre, form and rhyme. They are literalised as the language of the poem, an attempted alchemy at a basic level turning letters into verse into life. They are the building blocks of poetry here, mutating into metric feet and meaning. The verse is never *actually* DNA, demonstrating the gap between the physical and the imagined, language and the world. DNA is a similar means of describing the world to poetry, simply one type of version, a kind of aesthetics. The DNA verse here is also described as a 'litany', meaning both a set of repetitions and a church ritual. Donne was both a fleshy poet and Dean of St. Paul's, someone who melded the bodily and the divine. Roberts's presentation of the body is as a ritualised semi-divine incantation, something that might enable access to God.

The use of Donne as a key intertext here means that Symons Roberts's poem shares a motif with Duffy and Kunial, an engagement with a forbear within a poetics that conceives of legitimacy, inheritance, the shift from then to now. Christian Bök's *Xenotext* engages with Virgil but also presents in sonnet form – despite itself, this challenging verse still cleaves to shape. The importance of the relationship to a 'master' poet is common in writing on poetry, from T.S. Eliot to Harold Bloom, and these authors are clearly aware of what they are doing in setting themselves up against key figures. Each poet redefines a relationship with the past by emphasising the modernity of the now. They conceive of an interrelationship between then and now, understanding that DNA is something slippery yet material, relational yet unimportant. DNA is a new way of understanding the human in the present, and hence it changes our relationship to the past. Each writer shows this shifting in their mobilisation of old forms or use of old poets. Contemporary writing must re-see the human and interrogate modes of discourse about the species. These writers reach to poets who are pre-modern and pre-enlightenment, suggestively ignoring the *becoming* of the modern subject in the period 1700-1900 (as it is often understood theoretically), looking to forms and writers that are particularly historical yet constantly being reworked. The poets look to a particular period when the description of the human seemed to be reaching something new, something informed by empire and medicine and faith; but they pass over the development of 'modernity', and, particular, a kind of medicalised knowing of the body that Foucault and others have identified as occurring from the early nineteenth-century onward. These writers rest their practice in conceiving of the human as both historically contingent and specific, and also somehow attached to a species or a family that is older and stranger. They celebrate a transmission between then and now. As Bruno Latour argues, accentuating the complex mix of the human culture that 'has never been modern': 'Some of my genes are 500 million years old, others 3 million, others 100,000 years, and my habits range in age from a few days to several thousand years […] it is this exchange that defines us' (p. 75). Poetry understands this consonance and difference, the resonance of the then in the now. These poems are modern but archaic, much as the human is understood as a modern subject now but shares form with those from the past. Our DNA has barely changed over hundreds of years, but the ways in which we understand ourselves and *see* ourselves definitely have. 'Realism' is a historically contingent conception. These poets interrogate 'history' and its implications as a discourse, and they use DNA as a way of troubling categories of the real and of historical narratives.

Structural racism and identity: Lizzo, Residente and Kendrick Lamar

Aesthetic work made in response to consumer DNA tests is increasing as artists begin to establish a set of tropes and critical positions regarding the tests. The artists

David Blandy and Larry Achiampong produced an intervention for 'Trust me I'm an artist', part of Arts Catalyst's series *Dreamed Native Ancestry (DNA)*. Blandy and Achiampong undertook a series of Ancestry DNA tests to interrogate the claims made on their behalf, particularly around identity, and ask 'how does this removal of identity from its narrative and social dimension impact on understandings of race and relationships?'[51] Their work suggests a critical interest in the consequences of increasing home testing and database expansion, and particularly about bioethics. DNA testing here allows a wider set of discussions about ethics, politics, migration, biopolitics, and identity, and in turn suggests an engagement with the postgenomic condition.

The American singer and rapper Lizzo's 'Truth Hurts' (2018) includes the key line 'I just took a DNA test, turns out I'm 100% that bitch'. Later in the verse Lizzo differentiates between the 'human' and the 'goddess in me', splicing genetic identity with new feminist language of empowerment. Lizzo has trademarked the line '100% that bitch' for clothing and promotional purposes, and it speaks clearly to the singer's high profile regarding female self-confidence. After release of the track a publicity website allowed users to test themselves what percentage 'that bitch' they were, according to the 'Lizzo DNA Test'. The purpose of the test is to 'own' yourself, particularly in rejecting male arrogance; users shared their results on social media. The meme created discussion about genetic makeup and behaviour. The whole multi-spectrum event, happening across various platforms (Twitter, Instagram, Spotify, TikTok) and creating interactive discussion and content, demonstrates the simultaneity and complexity of contemporary media moments.

Whilst it was humorous the social media discussion focused on ideas of empowerment and confidence as being somehow part of a fundamental physiological identity. Lizzo's line responds to the phenomenon of widespread DNA testing, particularly amongst African American communities, but playfully subverts the findings as being about the 'now' rather than worrying about the past. It also comments upon the essentialism evident in the marketing of most DTCGT, insofar as the joke undermines the relatively spurious percentage claims of such firms and the wider idea that DNA can explain behaviour. The overlaying of notions of self-sufficient ownership with concepts of 'owning' property responds to concerns about genetic patenting and the commodification of genetic materials (see chapter X). The moment demonstrates clearly popular clichés about DNA as deterministic of behaviour. It shows, moreover, the increasing profile of DNA testing as a means for understanding identity. Lizzo's phrasing – 'turns out' – hints at the language of revelation that such testing carries popularly. The 'Lizzo DNA test' cheekily subverts the language of the major companies in outlining identity, whilst suggesting that looking to inside biologically might be a way of explaining and articulating identity.

Similarly, the Puerto Rican rapper Residente conceives of genetics as something that opens up fragile new possibilities within the self. Residente took a DNA test in 2015 that led him to reflect upon the diversity and hitherto unseen aspects within himself. The entire album (and documentary) Residente made after having his DNA tested for ethnicity in 2015 attempted to reflect upon the diversity and plurality he discovered therein: 'A little blood, a sample of saliva/ Send it to the lab and get it back and see them try to declassify/ A deoxyribonucleic gas just from one fiber'.[52] Residente conceives of his genetic make-up in ways that are fuelled by consumer technologies and contemporary understanding of the genealogical significance of biology: 'Scientists can separate a strand/ Tell you in percentages descendancies you long to understand' (ll. 17-18). Rapping with his cousin, Lin-Manuel Miranda, Residente switches between Spanish and English in his reflection upon how new genetic techniques enable the individual to conceive of themselves as the world in microcosm. Miranda is the writer of the musical *Hamilton* (2015), a world-famous reflection upon the lack of diversity in most American national history and the ways it might instead be told.[53] His involvement in the track overlays family and kinship community, but also introduces a pointed politics about the representation of the national past and the involvement of minority communities in that narrative (Figure 5.1).

This track's title, 'ADN/DNA', presents genetic coding as split between English and Spanish, multilingual, binary in naming, itself a double structure never touching but providing the building blocks for life. Residente suggests that such mathematical 'percentages' contain 'descendancies you long to understand', highlighting the affective impact that genetic testing might have on the participant. The track considers ethnicity and community, family and kinship, but also recognises how 'you'll find what you planned isn't quite what you get'. The track

FIGURE 5.1 Residente performing in Panamá, 2019

develops a model of innovation that has been allowed and enabled by genetic insight: 'And yet, you would make music/ Make something new from the rhythms and the fuses of chords beyond the images in the news' he raps, concluding 'If you can produce it your dreams are lucid, then you spit' (ll. 13-15).

Here Residente makes a joke having to 'produce' 'spit' – a term used for rapping (spitting verse) but also the source of the cells taken from his body to sample. His 'spit' contains within it all the multiple elements of his identity and his culture, and the new technology has allowed a revelation: 'you would make music/ Make something new'. This is a further example of poets and rappers reflecting upon the formal aspects of their art and aligning it with the work of genetic understanding. Similarly to the discussion of prosody above, Residente's use of 'spit' allows him to conceive of DNA as imbricated with creative work, pointing to new possibilities and new forms. These artists then use the genetic data to engage with their past and from this encounter to bring forth innovative work, part of the creative process whilst refusing to become subservient to the data. They are not inflected by their genetics but rather make the possibilities work for them: 'Send you to lands of ice dirt and sand/ A map of the world in the lines of your hand' (ll. 19-20).

This productive engagement can be seen clearly in Kendrick Lamar's track 'DNA', in which the rapper utilises the possibilities of postgenomic vision to specifically attack structural racism.[54] Lamar reaches for the metaphor of DNA to express something that is seemingly particular whilst also being elastic in meaning. Lamar gives voice to a popular belief that his genetic coding holds *within* it something innate that relates to his identity, whilst also arguing that he is able to contain multiplicity. His DNA contains history and culture as well as emotion and ability. He recognises a kind of genetic realism – this is *him* – but also seeks to disavow it, as his background is a complex multiplicity of influences and events, many of which have been imposed upon him and his community by a racist country. For Lamar, DNA is foundational, the building blocks of his physical identity but also the basis of his verse: 'I got, I got, I got, I got Loyalty, got royalty inside my DNA.'[55]

The track emphasizes the contradictions of his life: 'Cocaine quarter piece, got war and peace inside my DNA/ I got power, poison, pain and joy inside my DNA' (ll. 3-4). The acronym 'DNA' here becomes part of the punctuation, enabling the structure of the track. It drives the verse, and Lamar, forward. At the same time, it draws him back into his past, and the repetition of 'DNA' as a rhyme word means the verse itself continually echoes earlier lines. DNA is simultaneously *now* and resonating with *then*. It is multiple, diverse, complicated, temporally disjointed, both precise and expansive: 'I got hustle though, ambition, flow, inside my DNA' (l. 5) (Figure 5.2).

In his verse Lamar updates Walt Whitman's famous line 'I am large, I contain multitudes', outlining the contestation within himself and elements he may not control: 'I got dark, I got evil, that rot inside my DNA' (l. 13).[56] This inhabiting and rewriting of possibly the most famous line of American poetry signals a contestation of national identity enabled through genetics. Like the other writers considered here Lamar engages with his poetic inheritance and disrupts it. He also stages an encounter that seems to allow for continuum and rupture at the same

FIGURE 5.2 Kendrick Lamar, 2013

Creative Commons Attribution 3.0 Unported Licence, https://commons.wikimedia.org/wiki/File:Kendrick-lamar-1360479601.jpg.

Credit: Merlijn Hoek.

time. Lamar's African American experience is rooted in the Civil Wars that made Whitman famous. He presents a type of innate experience that is encoded somehow within but might also enable certain aspects of his character and career (hustle, ambition, flow). 'I was born like this [...] born inside the beast' (l. 6, l. 34), he suggests, claiming community and individuality simultaneously. His DNA forms him, makes him, is inescapable, but is also something that he claims, he owns, he inhabits. The beast is humanity, American society, race, all at once.[57]

The history of Lamar's family and the history of his community are bound together, with is account of how 'This is Paula's oldest son/ I know murder, conviction' (ll. 28-9). The context of his life is 'Burners, boosters, burglars, ballers, dead, redemption/ Scholars, fathers dead with kids' (ll. 30-31). This section ends with an unmistakeable threat: 'And I wish I was fed forgiveness' (l. 32). The sense that what has been done to me and mine has crept into our molecular structure – and we will revenge – plays on ideas common to epigenetics that genes can evolve in response to great trauma.[58] Lamar reflects the complex engagement of the African American community with DNA and genetic science.[59] For Lamar, then, his DNA is something to be proud of but also to fear, something that – despite the track's driving velocity – might drag you back, trip you up, open up things within that you may wish to ignore or avoid, that might be more truly *you* than you are happy to admit.

Lamar's and Residente's wide-selling and widely-viewed tracks are part of an ongoing complex imagining of genetics in popular culture. They are highly influential and reflect a complexity of understanding of how genetics create a bridge between then and now, a communal past and an individual in the present. Both tracks reflect a concern with identity, with ethnicity and race, with ability and legacy. Both see, as with the other writers considered here, the opportunity but also the threat of being postgenomic. Both engage aesthetically with the possibilities whilst celebrating the messiness. These poets and rappers work within Ato Quayson's definition of a 'history' that is self-conscious about its biases and implications:

> I ask for a history that deliberately makes visible, within the very structure of its narrative forms, its own repressive strategies and practices, the part it plays in collusion with the narratives of citizenship in assimilating to the projects of the modern state all other possibilities of human solidarity.[60]

The artists interrogate 'history' and its implications as a discourse, and they use DNA as a way of troubling categories of the real and of historical 'narrative forms'. In particular these texts offer, through their engagement with the past and consideration of a new relationship to history navigable through genetics, critique of scientific progressivism. In their consideration of the new types of genetic realism seemingly on offer, the poems and tracks reconceive of the human in history. On the one hand, being postgenomic seems to offer a kind of connectivity, a way of understanding the *human* within time and our relation to the past in new and innovative fashion. It presents an interrogation of older discourses of realism, of bodily completeness, of history and of race. On the other, writers increasingly see postgenomic discourses as a way of attempting to control, to impose, and close down meaning. In contrast, they celebrate fragmentation and strangeness, disintegration and metaphor, and highlight throughout that language – whether it be genetic data, or mathematics, allusion or motif – is a slippery, strange thing to be celebrated and mistrusted. They reach for the past as a way of undertaking this, as the postgenomic looks to the pregenomic in order to consider the continuum between each period. Each writer here redefines a relationship with the past by emphasising the modernity of the now. They conceive of an interrelationship between then and now, understanding that DNA is something slippery yet material, relational yet un-important. DNA is a new way of understanding the human in the present, and hence it changes our relationship to the past. Each writer shows this shifting in their mobilisation of old forms or use of old poets. Contemporary writing must re-see the human and interrogate modes of discourse about the species. These examples make the postgenomic as much as being made by it, and in their space of negotiation and innovation demonstrate the ways in which we might resist hegemony of all kinds. Instead these artists yoke new and old together, remaking the world and asking for a political aesthetics that is aware of itself and committed to a different future, as Residente argues: 'To remind you to struggle and strive that the trouble with life is/ that life isn't done with you yet'.

Accessing the Past and Changing the Future

Genetic engineering has been a staple of science and fantasy fiction, ranging from *Gattaca* (Andrew Niccol, 1997) to *Captain America: The First Avenger* (Joe Johnston, 2011).[61] Indeed, a large number of the hugely successful recent films based on comics depend on central characters who have had their genetic code changed in some way (Hulk, Spiderman, Captain Marvel, Ms Marvel, Luke Cage), reflecting contemporary anxieties about enhancement of human beings and also of the possibilities of genetic mutation (see the discussion of *X-Men*, below).[62] This concluding section looks at two important examples where genetic identity becomes imbricated in ways of challenging the past and a means for positing alternative pasts and futures.

Genetic connection: Assassin's Creed

Assassin's Creed (Ubisoft, 2007-) is one of the most popular games of all time, with global sales figures of the collected titles reaching over 140 million units.[63] It was made into a major film in 2016, demonstrating its range and appeal as a text. The game has attracted a lot of attention from historical scholars, mainly because it introduces educational re-enactment into a highly popular gaming format.[64] In the game the player controls an avatar-character who explores a particular historical locale (these have included Paris in 1789, London in 1868, Rome in the late fifteenth-century, the Peloponnesian war). The location is highly significant to the story and gameplay. The historical setting is faithfully rendered, with relatively accurate maps, buildings, and characters (including appearances from figures such as Herodotus, Da Vinci, Marx and Queen Victoria). Therefore, the player immerses themselves in the environment and explores it, developing a different engagement with pastness and a new means for understanding historical experience. Whilst it is primarily a combat game, more recent iterations have emphasised the educational experience, with scientific and cultural missions, expanded encyclopaedias, and an even greater emphasis on historical significance and accuracy. Scholars have long celebrated the way that the complexity of the texts offer an opportunity to evaluate 'processes of cultural memory' (Hammar, p. 377), whilst also recognising the series' continuing creative use of anachronism.[65] *Assassin's Creed* is a significant resource for popular historiography, enabling a player to understand the past in different and bodily fashion, and developing models of re-enactment practice and immersive engagement.[66] It adds 'play' to the experience of the past, giving the participant relative freedom to explore and understand in ways that bypass the textual.[67]

One central aspect that has been relatively under considered in the approaches to the text, however, is the way that it utilises genetics as its primary means for engagement with the past. The text presents a complex way of conceptualising DNA and its ability to connect us with history, and hence suggests an important way that this genetic continuum has been imagined over the past two decades.

The game presents genetics as a way of directly accessing the past, and as such is an important text for conceptualising how genetic code works and how it enables engagement with both personal and collective histories. The first five titles focused on the character of Desmond Miles who is a descendant of a number of 'assassins' throughout history. Miles is kidnapped by a company called Abstergo who force him to enter a 'science fiction genetic time machine' called the Animus.[68] The technology in the Animus allows Miles to experience part of the lives of his forebears. Their memories are *stored* in his DNA, and the machine accesses them through the genetic material and creates a simulation which the player engages with through avatars. Therefore, the crucial link between then and now is genetic. DNA allows a kind of time-travelling link to past memory and experience, and, importantly, the player participates in this imaginative movement.

Miles's ancestors include an Anglo-Kanien'kehá, a Syrian, an Italian, and a Welshman. The complex hybridity of his ethnic and genetic background demonstrates the games' commitment to a type of performative diversity in engaging with the past but also shows an understanding of genetic complexity in terms of inheritance. The seeming aspirational complexity of the postgenomic imaginary is here utilised within an economic nexus. The games have often sought to appear to decentre normative histories: the first game in 2007 featured a Syrian protagonist only years after 9/11; *Assassin's Creed IV: Black Flag* includes a long section in which the player controls Adéwalé, an African-heritage assassin supporting the slave uprising in Haiti in 1715. Indeed, the wider game context is an ongoing battle between the Assassins and the Templars, suggesting a conspiracy version of history in which the shadowy Templars have been running society from the shadows for centuries. In this revisionist vein the emphasis of bodily memorial connection to the past, rather than textual, figures a challenge to standardised ways of remembering and of accessing history.[69] In various of the games there are interzones, lobbies between the historical simulation and the 'real' contemporary world, and these emphasise the genetic connection with molecular models and double helixes floating around. Furthermore, the simulation often fails, or glitches, and there are also simultaneous events happening in the contemporary world that the player needs to engage with. Formally, then, the game continually reminds the player that this is a simulation enabled by a genetic connection to the past. Memory here is something that can be accessed through genetics; it is something buried in our bodies that we do not comprehend or acknowledge. The right technology can connect us therefore to multiple pasts and let us 'live' them again through the connection of our DNA. The game structure demands that a player 'synchronise' the memory (that is, complete the mission) and this is illustrated in a percentage as part of a DNA strand. A player who is killed is 'Desynchronised' and must begin again. In this representational strategy of the game, therefore, DNA is conceptualised as literally storing historical information and knowledge.

Assassin's Creed directly links genetics, historical experience, play, and bodily understanding of the past. DNA is a means for entering into and replaying the memory of those who came before. Genetic links with the past can be mobilised to enable a re-enactment and an immersive historical experience. The motif here is of DNA as a strand of connection, directly linking then and now. The suggestion is that each of us has, latent in their cells, such links to the past. History literally lives in us, and we are both the sum of all its parts and unique. Our DNA is made up, in this reading, of all the experiences of our ancestors. Somehow these events can be made to live again, in a full-spectrum, sensory bodily experience.[70] Within our DNA is the information we might need to access the past, to bodily engage with history, and to re-enact events.[71] The physical re-inhabiting of past moments through genetic code enables a re-enactment to happen (although on the games terms this is simply an 'enactment', a witnessing of what happened). The game suggests a continuum between then and now, a traversable link made physical in the strands of our DNA. The game itself is a visualization of this traversing, moving between then and now via genetic code. *Assassin's Creed* posits that we can use our genetic information to time-travel, and that deep within us is the historical truth. Furthermore, the meta-avatar gameplay – we play a game simulation as Desmond who in turn enters a simulation to 'play' as an ancestor – likens this genetic connection to our own engagement with the past through the game. Players surf Desmond's DNA to get to the past and explore it in a way which challenges traditional modes of understanding. Importantly, the interface with the past is ludic, something to be enjoyed and played with, rather than directional and textual. Education here, such as it is, seems self-directed and unexpected. The player fights through history rather than 'learning', a telling of the past which emphasises combat and violence to be the means of achieving progress. The game, then, figures genetics as something that enables connection to the past and allows a re-enactive experience of history. Rather than a dreamed, aspirational, or imagined link to the past, DNA here allows a proleptic haunting to happen, as ghosts from the future stalk the past.

Genetic disruption: X-Men

The successful and influential *X-Men* franchise of films (13 to date) began in 2000 and kicked off an explosion in comic-book films that has lasted two decades.[72] The films are based on an influential comic book series that has engaged with notions of otherness, queerness, and difference.[73] Expanding upon the premise of the long-standing comic series (begun in 1963), the films depend on the idea that DNA can evolve through mutation and work through the implications of there being a 'new' and different species of human on the planet. The films also demonstrate the anxiety attendant upon *being* postgenomic, worries about how data can be understood about us, and how the 'story' in our genes might change us without our knowledge.[74] The *X-Men* series articulates a palpable anxiety about genetic make-up, aligned with an imperfect acculturated understanding of key concepts (mutation, inheritance, ethnicity). Across 20 years they have mirrored the shifts in public understanding and public fear relating to genetics.[75] The films are a resource that illustrate how Western society post-2000/ post-Human Genome Project has been consumed with concerns about what our new genomic knowledge might affect: challenges to structures of race, sexuality, disability; increased genetic engineering; genetic 'cures'; problematic bioethics; complicated genetic inheritance.

The engagement of modern mainstream superhero films with the past is patchy. *Captain America: The First Avenger* (Joe Johnston, 2011) set in the 1940s, and *Captain Marvel* (Anna Boden and Ryan Fleck, 2019), set in the 1980s, are the only major Marvel films to be entirely set in precise historical locations, whilst both DC *Wonder Woman* films (Patty Jenkins 2017, 2020) similarly situate themselves in the relatively familiar past (the 1940s and the 1980s). Neither set of films engages with the problematic ethics of imagining enhanced and superhuman figures in historically specific periods, although the *Captain America* films do reflect upon the revenant superhero figure out of time. There has been no sustained conceptualisation of how the superhero film genre might engage with the costume drama formats, and a generic aesthetic engagement with the past is generally lacking. That said, increasingly superhero films and TV series are working through issues such as transhistorical pastiche (*WandaVision*, 2021), legacies of colonialism (*Black Panther*, 2020), time travel (*Loki*, 2021), the ethics of non-human identity (*Wonder Woman: 1984*, 2020), Ostalgie (*Black* Widow, 2021) and temporal grievability (*Avengers: Endgame*, 2019). These films' (and their television counterparts) almost obsessive concern with temporality, rather than historicity, is important to note.

In contrast, the *X-Men* films have actively engaged with historical representation, particularly through a carefully developed sub-series beginning with *X-Men: First Class* (Matthew Vaughn, 2011), set in 1962. Whilst the original set of *X-Men* films (2000-2006) are carefully modern, they do contain moments of reflection upon the past, most importantly at the opening of the first movie when the mutant Magneto is taken to Auschwitz as a young boy. This powerful scene begins the film series by locating the historical trauma and suggesting that this violence led to destruction in the contemporary moment.[76] The sub-series of films explores this originary trauma, not least through recording the Nazi experimentation on Magneto during his capture and his subsequent revenge. The films, therefore, have been concerned throughout with conceptualising what it means to be super-, supra-, and non-human within recognisably historical moments, and in riffing upon what this might mean. In particular, they have situated contemporary genetic anxieties about humanness in the postgenomic moment within historical locations, placing characters who are post-human in their genetic mutation in a pre-genomic moment.

This chronological fluidity is important to their iterative power as films exploring types of humanness and allows the texts to interrogate normative assumptions about genetically-inflected identity by historically displacing the discourse. *Days of Future Past* (2014), the most recent film in this sub-series, combines an understanding of genetic anxiety with a reflection upon historical fictiveness which links the flux of DNA with the malleability of temporality. As such it articulates the 'flux' of double helix history discussed in the Introduction. Whilst the X-Men films have been concerned with genetics from the outset, with the first movie coming in 2000 as the human genome was being sequenced, *DFP* folds this interest into a historical narrative.

The series it is in is a sub-set of the X-Men franchise taking place in the past, and hence the imbrication of genetics and past representation is important to consider. The scientific premise of *Days of Future Past* is that the authorities will use the DNA of a mutant against them, itself something seemingly beyond the pale: 'I never imagined they'd use Raven's DNA to do it' says the usually cynical Magneto. The historical premise is that the mutant Wolverine has been sent back in time to the 1960s to prevent this, and to save the future. The film articulates anxieties surrounding genetic science within a historical narrative. Additionally, the film challenges linearity and temporal order, suggesting that they, like genetics, can be challenged, circumvented, and reconfigured. As it plays around with time and our experience, the film suggests that this is only possible because our understanding of the human has been itself disrupted by genomic science. Stability and rationality, represented as bodily and species normativity and temporal order, are no longer achievable.

The film plays on anxieties relating to genetic mutation and aligns these with the malleability of time and history. Actual events – the assassination of Kennedy, for instance – are given a new resonance or meaning, made strange through their association with the 'new' *Homo superior*. The mutants stand in for a new way of understanding the species. They allow us to think about the human mutated, changed, and shifted due to genetic evolution. This mutation has concomitant impact upon the nature of time. The series of films posited an alternative timeline to the 'normative' time of the canonical *X-Men* series. However, *DOFP* adds an alternative timeline to this, as Wolverine is sent back in time to change historical events in the 1960s and prevent the development of a particular type of mutant-killing sentinel. History, already disrupted, is again divided and rethought. The genetic shift into the metahuman/ mutant human puts time out of joint. The humans recognise this, weaponising the DNA of the mutants in an attempt to eradicate their radical possibility and the threat they represent to rationality and human wholeness. The war-machine sentinels created at the end of the film target the 'Mutant X-gene, a genetic guidance system that can lock on to from half a mile away and won't fire until it has identified the target'. For Magento, this demonstrates the continuing foundational evil of *Homo sapiens*, multiplied by the new genetic understanding. Genetic knowledge here simply feeds human destructiveness, enabling them to resituate the normative and to heal time through committing genocide. The film presents a bleak vision of humanness as destructive and vicious in the face of challenge. Rather than conceptualise reconfigured genetic identity as something compelling and liberating, the films suggest that the more that is known about DNA, the more destructive the consequences might be.

Notes

1 See, for instance, Jackie Stacey, *The Cinematic Life of the Gene* (Durham, NC: Duke University Press, 2010).
2 See Jonathan Roberts, Louise Archer, Jennifer DeWitt and Anna Middleton, 'Popular culture and genetics; friend, foe, or something more complex?', *European Journal of Medical Genetics* 62:5 (2019), 368–75.

3 See discussions in Josie Gill, *Biofictions* (London: Bloomsbury, 2020), Lara Choksey, *Narrative in the Age of the Genome* (London: Bloomsbury, 2021), Clare Hanson, *Genetics and the Literary Imagination* (Oxford: Oxford University Press, 2020), the special issue edited by Mandy Bloomfield, and Clare Hanson 'Beyond the Gene: Epigenetic Science in 21st Century Culture' *Textual Practice* 29:3 (2015) and the special issue of *Medical Humanities* 47:2 (2021) edited by Clare Barker on 'Global Genetic Fictions'.

4 *Biofictions*, p. 5.

5 *Narrative in the Age of the Genome*, p. 7.

6 Quoted in Jamie Condliffe, 'Cryptic poetry written in a microbe's DNA', *New Scientist*, 4 May 2011, https://www.newscientist.com/blogs/culturelab/2011/05/christian-boks-dynamic-dna-poetry.html [accessed 26 April 2018].

7 Isabel Waidner, 'Christian Bök's *Xenotext Experiment*, Conceptual Writing and the Subject-of-No-Subjectivity: "Pink Faeries and Gaudy Baubles"', *Configurations* 26 (2018), 27–46.

8 NPG Archives 46/66/66 – RP 6591–6592 Registered Packet Sulston, Second Folder: NPG 6591 Sir John Sulston, Memorandum of Commission, 18 November 2001.

9 Ken Arnold, "Marc Quinn," *Tate* (Spring 2002), 19.

10 NPG Archives 46/66/66 – RP 6591–6592 Registered Packet Sulston, Second Folder: NPG 6591 Sir John Sulston, Letter Tim Moreton to John Sulston, 25 September, 2006.

11 NPG Archives 46/66/66 – RP 6591–6592 Registered Packet Sulston, Envelope 4 'THE WELLCOME TRUST NPG 6591, 6592(1–2), Pre-meeting Minutes of Wellcome Scientist Meeting, 6 March 2001, point 1.2.

12 NPG Archives 46/66/66 – RP 6591–6592 Registered Packet Sulston, First Folder: Conservation Record: Sulston, 'Conservation Report August 2004'.

13 NPG Archives 46/66/66 – RP 6591–6592 Registered Packet Sulston, First Folder: Conservation Record: Sulston, 'Conservation Report August 2004'.

14 NPG Archives 46/66/66 – RP 6591–6592 Registered Packet Sulston, First Folder: Conservation Record: Sulston, Email Kathleen Soriano to Ken Arnold, 6 April 2004, 09:09.

15 NPG Archives 46/66/66 – RP 6591–6592 Registered Packet Sulston, First Folder: Conservation Record: Sulston, 'Conservation Report August 2004'.

16 NPG Archives 46/66/66 – RP 6591–6592 Registered Packet Sulston, First Folder: Conservation Record: Sulston, 'Conservation Report 9 November 2016', p. 1.

17 NPG Archives 46/66/66 – RP 6591–6592 Registered Packet Sulston, First Folder: Conservation Record: Sulston, 'Conservation Report August 2004'.

18 NPG Archives 46/66/66 – RP 6591–6592 Registered Packet Sulston, First Folder: Conservation Record: Sulston, 'Conservation Report NPG 6591 (+ spares) October 2012', p. 1.

19 NPG Archives 46/66/66 – RP 6591-6592 Registered Packet Sulston, First Folder: Conservation Record: Sulston, 'Conservation Report 9 November 2016', p. 2.

20 NPG Archives 46/66/66 – RP 6591-6592 Registered Packet Sulston, Fourth Folder: 'THE WELLCOME TRUST NPG 6591, 6592(1–2) 'Pre-meeting Minutes', Wellcome Scientist Meeting, 6 March 2001. Also 'Meeting Minutes', 6 March 2001 of 'Wellcome Scientist Meeting' includes '1.4 Likely that object will need renewal but **JS** [John Sulston] and Marc Quinn to discuss production and determine how long term preservation of prime object is possible', NPG Archives 46/66/66 – RP 6591-6592 Registered Packet Sulston, Third Folder, Minutes.

21 NPG Archives 46/66/66 – RP 6591–6592 Registered Packet Sulston, First Folder: Conservation Record: Sulston, Email Kathleen Soriano to Ken Arnold, 6 April 2004, 09:09

22 NPG Archives 46/66/66 – RP 6591–6592 Registered Packet Sulston, First Folder: Conservation Record: Sulston, Email John Sulston to Kathleen Soriano and Ken Arnold, 04 May 2004, 19:56.

23 NPG Archives 46/66/66 – RP 6591–6592 Registered Packet Sulston, First Folder: Conservation Record: Sulston, Email John Sulston to Kathleen Soriano and Ken Arnold, 04 May 2004, 19:56.

24 NPG Archives 46/66/66 – RP 6591–6592 Registered Packet Sulston, First Folder: Conservation Record: Sulston, Email Kathleen Soriano to Ken Arnold, 6 April 2004, 09:09

25 NPG Archives 46/66/66 – RP 6591–6592 Registered Packet Sulston, First Folder: Conservation Record: Sulston, Email Ken Arnold to Kathleen Soriano, 19 May 2004 19:35.

26 NPG Archives 46/66/66 – RP 6591–6592 Registered Packet Sulston, Second Folder: NPG 6591 Sir John Sulston, Email Bronwyn Terrill to Clementine Hampshire, 6 September 2010, 11:24; Email Kathleen Soriano to Ken Arnold, 16 April 2010, 10:53.

27 NPG Archives 46/66/66 – RP 6591–6592 Registered Packet Sulston, Second Folder: NPG 6591 Sir John Sulston, Emails Bronwyn Terrill to Juliet Simpson, 12 July 2010 12:52.

28 NPG Archives 46/66/66 – RP 6591–6592 Registered Packet Sulston, Second Folder: NPG 6591 Sir John Sulston, Fay Blanchard to Ken Arnold, 24 May 2013, 16:19.

29 NPG Archives 46/66/66 – RP 6591–6592 Registered Packet Sulston, Second Folder: NPG 6591 Sir John Sulston, Fay Blanchard to Tom Ziessen and Ken Arnold, 22 July 2013, 15:04.

30 NPG Archives 46/66/66 – RP 6591–6592 Registered Packet Sulston, Second Folder: NPG 6591 Sir John Sulston, Typescript of comments that eventually became part of the exhibition press pack.

31 'Portrait of the Week', *The Guardian*, September 22, 2001, https://www.theguardian.com/culture/2001/sep/22/art [accessed 15 December 2017].

32 Susan Merrill, *Epigenetic Landscapes: Drawings as Metaphor* (Durham, NC: Duke University Press, 2017).

33 *How to be both* (London: Hamish Hamilton, 2014).

34 *Girl, Woman, Other* (London: Hamish Hamilton, 2019).

35 See Gill, *Biofictions*, pp. 101–21.

36 *Poetry* 111:3, December (1968) p. 175.

37 Carolyn Dinshaw, 'Temporalities' in Paul Strohm, ed., *21st Century Approaches: Medieval* (Oxford: Oxford University Press, 2007), pp.107–23 (p. 109).

38 See Judith Roof, *The Poetics of DNA* (Minneapolis and London: University of Minnesota Press, 2007).

39 Hannah Sullivan, "The Sandpit after Rain," in *Three Poems* (London: Faber and Faber, 2018), 4:3.

40 Zaffar Kunial, "Self Portrait as Bottom," in *Us* (London: Faber and Faber, 2018), 29–30 (ll. 19, 20).

41 'Self-portraits in poetry are legion' writes Eloisa Amezuca introducing Rita Dove's 'Self-Portrait' in *The New York Times*, 1 November 2018, https://www.nytimes.com/2018/11/01/magazine/poem-self-portrait.html.

42 In *Self-portrait in a convex mirror* (Manchester: Carcanet 2007), p.74, first published in 1975. The opening lines run 'As Parmigianino did it, the right hand/ Bigger than the head, thrust at the viewer', referring to Parmigianino, 'Self-portrait in a convex mirror' (1524), Kunsthistorisches Museum, Vienna, Austria.

43 Walter Benjamin, "The Task of the Translator," trans. Harry Zohn, in *Selected Writings, Volume 1: 1913–1926*, ed. Marcus Bullock and Michael W. Jennings (Cambridge, MA: Harvard University Press, 1996), 253–63 (p. 254).

44 Benjamin, "The Task of the Translator," 254.

45 See the discussion of contemporary art, race, and genomics in Alys Eve Weinbaum, 'Racial Aura: Walter Benjamin and the Work of Art in a Biotechnological Age', *Literature and Medicine* 26:1 (2007), 207–39.

46 This is due to the fact that the database is mainly made up of customers from those locations, given the ways that DTC companies have developed.

47 *Twelfth Night* II.i.176.

48 Michael Symons Roberts, 'To John Donne' in *Selected Poems* (London: Cape, 2016), p. 62.

49 Haraway, *Modest Witness*, 64.

50 The shift from the clinical to the molecular gaze is outlined by Nikolas Rose, *The Politics of Life Itself: Biomedicine, Power, and Subjectivity in the 21st Century* (Princeton, NJ: Princeton University Press, 2006).

51 'Trust me I'm an Artist: DNA Ancestry Testing With Larry Achiampong and David Blandy', 8 November 2017, http://trustmeimanartist.eu/events/dna-ancestry-testing-larry-achiampong-david-blandy/.

52 'AND/DNA' from *Residente* (Fusion Media Group, 2017), ll. 14–16.

53 Philip Gentry, '*Hamilton*'s Ghosts', *American Music* 35:2 (2017), 271–80.

54 On Lamar see the thoughtful discussions of his work (and particularly the *DAMN* album) in Christopher M. Driscoll, Monica R Miller and Anthony B. Pinn (eds.), *Kendrick Lamar and the Making of Black Meaning* (London and New York: Routledge, 2019).

55 'DNA' from *DAMN* (Top Dawg Entertainment, 2017)., ll. 1–2.

56 Whitman's line 'I am large, I contain multitudes' from his 1855 'Song of Myself' section 51, in *Leaves of Grass* (London: W.W. Norton and company, 2002).

57 See Paul Gilroy's discussion of Snoop Dogg's 'infrahumanity' in *Against Race* (Harvard, MA: Harvard University Press, 2000), pp. 201–6.

58 See for example Rachel Yehuda and Linda M. Bierer, 'The relevance of epigenetics to PTSD', *Journal of Trauma Stress* 22:5 (2009), 427–34.

59 Discussed in depth in Alondra Nelson, *The Social Life of DNA* (Boston, MA: Beacon Press, 2016).

60 Ato Quayson, *Postcolonialism: Theory, Practice, or Process?* (London: Polity Press, 2000), 48.

61 See Jackie Stacey, *The Cinematic Life of the Gene* (Durham and London. Duke University Press, 2010) and Priscilla Wald and Jay Clayton (eds), 'Special Issue: Genomics in Literature, Visual Arts, and Culture', *Literature and Medicine* 26:1 (2007).

62 Eugene Thacker, *The Global Genome: Biotechnology, Politics, and Culture* (Cambridge, MA: The MIT Press, 2005).

63 Ali Jones, 'Assassin's Creed all-time sales tops 140 million', *PC GamesN*, 27 September 2019, https://www.pcgamesn.com/assassins-creed-sales [accessed 29 June 2020].

64 See Douglas N. Dow, 'Historical Veneers: Anachronism, Simulation and Art History in *Assassin's Creed*' in *Playing with the Past* ed. Matthew Wilhelm Kappell and Andrew B.R. Elliot (London: Bloomsbury, 2013), pp. 215-32, and Emil Hammar, 'Counter-hegemonic commemorative play: marginalized pasts and the politics of memory in *Assassin's Creed: Freedom Cry*', *Rethinking History* 21:3 (2017), 372–95.

65 Christopher Leffler, 'Memory games: history, memory and anachronism in the Paris of *Assassin's Creed: Unity*', *Contemporary French Civilization* 44:1 (2019), 81–99.

66 Brian Rejack, 'Toward a virtual re-enactment of history: video games and the past', *Rethinking History* 11:3 (2007), 411–25.

67 See the discussion of this in Adam Chapman, *Digital Games as History* (London and New York: Routledge, 2016).

68 Chapman, *Digital Games*, p. 84 n. 15.

69 The queer, transgressive and transformative potential of gaming is discussed in Bonnie Ruberg, *Video games have always been queer* (New York, NY: NYU Press, 2019).

70 See Mary Flanagan, *Critical Play* (Cambridge, MA: MIT Press, 2009).

71 Christian Casey, '*Assassin's Creed Origins*: Video Games as Time Machines', *Near Eastern Archaeology* 84:1 (2021), https://doi.org/10.1086/713365.

72 Jeffrey A. Brown, *The Modern Superhero in Film and Television* (London and New York: Routledge, 2016).

73 See Alexandro Segade, 'X-Men', in *Keywords for Comics Studies* ed. Ramzi Fawaz, Shelley Streeby and Deborah Elizabeth Whaley (New York, NY: New York University Press,), available at https://keywords.nyupress.org/comics-studies/essay/x-men/, and Anthony Michael D'Agostino, '"Flesh-to-Flesh Contact": Marvel Comics' Rogue and the Queer Feminist Imagination', *American Literature* 90:2 (2018), 251–81.

74 See Melanie Kohnen, *Queer Representation, Visibility and Race in American Film and Television* (London and New York: Routledge, 2015), pp. 56–70.

75 See Geoffrey I. McFadden and Naja Later, 'Evolution: Of X-Cells and X-Men', *Current Biology* 27:11 (2017), R408–9 and the essays in Claudia Buccifero, ed., *The X-Men Films: A Cultural Analysis* (London: Rowman and Littlefield, 2016).

76 Cynthia D. Porter, 'Germans and Genes on Screen: Marvel's *X-Men* Films', *Journal of Literature and Science* 13:1–2 (forthcoming, 2021).

6

SELF

This final chapter illustrates the huge changes for 'amateur' historians in using DNA to understand their past. In particular, the chapter looks at the phenomenon of genetic genealogy, the use of DNA for family history. Genetic genealogy is one of the key ways that DNA is understood around the world, and in particular articulates the types of historical information that is understood to be available, and the questions that might be asked of it. The chapter focuses closely on the ways in which family historians use genetic data as a means for investigating and telling their past. It allows them new insight and structures their thoughts about the past and how it might be accessed. In doing so genetic data can seem to transform the relationship of the contemporary self to the past, and to emphasize an unseen, bodily connection. Commercial DNA work also raises questions about privacy, ownership, and the commodification of genetic data that all impact upon the models for, and opportunities to, investigate the past. An analysis of the recent phenomena of genetic genealogy documentaries allows us to see how this new way of accessing the past contributes to the popular historical imaginary. The chapter concludes with a last discussion of the revisionist practice of Henry Louis Gates, Jr., who utilises genetic information in a profoundly restitutive fashion. This final section outlines how historical information derived from genetic investigation might radically challenge normative ideas of selfhood, nation, and communal experience.

DNA Testing and Family History

Family historians are a large group of researchers that have often been marginalised by mainstream historical practice.[1] Their work is often characterised as 'amateur' and ignored as a valid historical mode. This is despite the fact that as a group, family historians are careful, resourceful, and increasingly well-trained.[2] Over the

DOI: 10.4324/9781003052975-7

past decade family history has become more recognised as an important historical approach, and the community of family historians around the world have been considered as an important repository of intergenerational knowledge and understanding of how historical process works.[3] This is part of a wider recognition of the importance of 'amateur' history-making.[4] Family historians are a great example of a self-sustaining community of historical practitioners, with extensive and elaborated support including national and local associations, extensive electronic resources and wide-ranging social media presence.[5] Family historians are highly innovative in terms of their work, early adopters as a community of Internet resources, social media, and other research tools. They keep and organise their own information, sometimes using commercial software.[6]

Whilst genealogy had been a relatively big area, after the global success of TV series such as *Who Do You Think You Are?* and the expansion of online databases, the field went through a significant boom in the early 2000s.[7] There are growing numbers of university-level courses catering to the booming market in family history, as well as numerous free-to-access Massive Open Online Courses (MOOCs) run by universities and genealogical societies. Family historians are also supported by several key companies or websites, including Ancestry, MyHeritage, Findmypast, and Family Search, who provide access to global databases of information. Through their online platforms these companies allow users to create family trees using particular software and to share their research with a wide network of like-minded genealogists. These providers allow access to census data, archives and libraries, and the family historian grows their family tree through engaging with other users and undertaking their own research. Indeed, Ancestry is now one of the biggest providers of historical data in the world, and its activities enable engagement with billions of historical documents to millions of people globally.[8]

Family historians had been offered DNA-derived services from around 2000, with a set of companies working on offering Y-Chromosome and mitochondrial (mtDNA) tests.[9] These were generally quite specialist products, allowing new kinship and research connections through direct male and female lines.[10] Y-chromosomes, passed from father to son, were looked at in combination with surname analysis.[11] Family historians had undertaken surname studies for years, considering the patterning of names geographically. Y-chromosome information seemed to suggest further connections between groups of men with the same surname; it allowed individuals to test how they were related to these groups and to expose new connections.[12] The BBC Radio series *Surnames, Genes, and Genealogy* (2001) outlined this work for a popular audience. Attention to DNA in relation to surnames provided a 'means to determine the genetic make-up of past populations' and to 'assess the impact of historical population movements'.[13] Commercially such work allowed companies to suggest common ancestors, and to enable family historians to confirm their research accordingly.

A second type of DNA genealogy work focussed on using Mitochondrial DNA to find connections to the direct maternal line. Such DNA changes very slowly over time so can suggest direct relationships with broad groups. This was the basis for the Oxford Ancestors company, set up by Bryan Sykes in 2000. The project connected users to wide groups, claiming that there were '7 daughters of Eve' living between 11000 and 45000 years ago, to which everyone on the planet was ultimately related.[14] This type of 'deep dive' work was popular and important in developing a sense of DNA identity in relation to deep time. It was part of a wider cultural acknowledgement of 'species' history and was important in locating genetic identity as something that might be shared over thousands of years.[15] The Oxford Ancestors commercial outputs situated DNA heritage as something that connected users to the ancient past, and considered them as part of 'humanity' more widely.

Mitochondrial and Y-Chromosome DNA tests became increasingly used by commercial companies offering clients a connection to their past. The first genealogical DNA company in the USA, FamilyTreeDNA (FTDNA), was founded in 2000 and offered a combination of the two approaches. The company was founded by Bennett Greenspan, who had seen the work on Sally Hemings's DNA (see Chapter 1) and 'wanted to know whether launching a genetic genealogical-testing company was a plausible idea'.[16] The company grew quickly, and by 2006 revenue was $12 million.[17] FamilyTreeDNA continues to offer Y-DNA and mtDNA tests enabling customers to 'Learn about your personal history and follow the path of your ancestors'.[18] Early genetic genealogy companies were particularly aimed at certain groups, including the Jewish community and African American community.[19] In 2007 FTDNA partnered with Henry Louis Gates, Jr. (see below), to launch AfricaDNA.com, a database aimed directly at African Americans.[20] Rick Kittles had set up African Ancestry in 2003 as a company concerned directly with tracing African descent. Alondra Nelson argues that such work can be key for development of ideas of community and self: 'test results are valuable to 'root-seekers' to the extent that they can be deployed in the construction of their individual and collective biographies'.[21] At the same time, though, Nelson warns that ancestry tests risk conferring inflexible templates of ethnicity and identity. The assumption that DNA confers ethnicity or that race is a fixed genetic characteristic has become increasingly assumed during the rise of online DNA companies (see Chapter 4, and later sections).

During the period 2006-09 services offering DNA-derived information about ancestry, and family history began to grow in number, scope, and visibility. Increasingly, though, it became clear that Autosomal DNA, which gives data about paternal and maternal ancestry, would give a more complex sense of genetic framework. This type of data would allow suggestions to be made about the percentage of DNA shared between populations. Combined with population information this allows a reading of genetic information and in particular allows the suggestion of ethnic background. Autosomal DNA is more flexible in its application, and hence has become the industry-standard way of approaching genetic genealogy. FTDNA began to offer autosomal testing in 2010. Since 2010 and the arrival of widespread, cheap, heavily marketed Direct To Consumer

Genetic Testing, family historians have been offered products from an increasing set of companies such as AncestryDNA, 23andMe, MyHeritage DNA, and FindmypastDNA. With the opening up of genetic technologies, and the lowering of cost of testing, the family history DNA market has become huge, and widespread.[22]

The Rise of the Genealogical Biotech Companies

It is important to see the development of family history DNA testing as inter-twined with the increase in 'leisure' DNA testing. One of the first major companies to see the potential market for leisure DNA testing was 23andMe.[23] Originally founded in 2006 by a group with strong links to Google, 23andMe sought to provide genetic information and interpretation to customers.[24] The company grew relatively quickly and were at the forefront of 'companies offering to unlock the genetic mysteries of your past (ancestry), present (human traits) and future (susceptibility to health conditions)'.[25] Writing about DTCGT in 2008, Hogarth, Javitt and Melzer argued that 'Despite the inherent fluidity of the marketplace, the commercial allure of DTC testing, coupled with the lack of regulatory barriers to market entry, has led to a steady stream of new entrants', noting that 'more than two dozen DTC companies exist worldwide'.[26] Key to the expansion of the companies was their online identity, and thus ability to engage customers in particular ways. DTGT services shift the emphasis of genetic research and consumption:

> the curious consumer becomes a client and a research subject. Companies using consumer samples and data to conduct research are in essence creating databases of information that can be mined and studied in the same way as biobanks and databases generated by academic institutions.[27]

What 23andMe recognised was that there was a huge market that would be in-terested in discovering their genetic make-up, and in return users would share their data and allow it to be manipulated by the company. Such companies have been criticised for not providing sufficient counselling.[28] Indeed, the success of Ancestry (see below) was predicated on the fact that it decided against offering health advice (after initially launching 'Ancestry Health'). Hence the company was not obliged to offer any further support to individual users. This meant it was much more agile as a company, although its product was not as widely applicable.

Genetic testing for family history is presented as a tool to help break down 'brick walls', leading to the solving of problems that might have persisted for decades. The databases for each company suggest connections between user in-formation and accordingly attempt to 'connect' customers to each other.[29] Family historians around the world use genetic information to understand their back-grounds in more depth and to connect with those who might be related to them. They also recognise that the information generated might severely change the way they think about their past, with unforeseen revelation increasingly common-place.[30] DTCGT services for family history and ancestry largely suggest for a user their ethnic make-up and geographical origins. For some users, this is not parti-cularly interesting.[31] That said, increasingly DNA testing for ancestry is marketed as 'lifestyle' genetics, allowing the user a sense of heritage connection to a com-munity or an historical way of understanding 'who they are'. The marketing for the major companies addresses novelty, revelation, and historical situatedness. Millions of leisure users have paid the companies to gain a sense of their ethnicity and connection to communities. The data produced is interpreted by the com-panies and communicated through their websites. These websites package the DNA data, presenting timelines, pie charts, graphs, maps and ethnicity estimates, all based on a reading of the user's genetic code.

The companies allow individual users to download their full dataset of raw DNA data. To 'read' and use this information family historians have created da-tabases and spreadsheets, written bespoke software or utilised sophisticated online crowd-sourced or community-curated programmes such as DNA Painter or the suite of facilities at GEDMatch.com. Tools enable the genetic data to be mined and interrogated. Such software includes GEDMatch's 'Lazarus' programme, which allows the user to postulate the genetic make-up of dead relatives, and 'Promethease', which allows family historians to have their genetic information read for its health implications. Some similar tools are provided by the big companies as well, keen that users manipulate data within their ambit. The family historian will organise their own data and may also manage the information of others. They will often have access to the genetic information of many other participants (voluntary given) and will be connected through the company in-terfaces to many possible 'matches'.

In 2012, Ancestry began to offer DNA testing as part of its suite of family history services (initially within the US, and then across the world after 2015). The service would allow for several things: a biogeographical report (often interpreted as an estimate of ethnicity); availability of some raw genetic data; and the link to a wider network of family historians all of whom had done the same test. Ancestry is a major influence on the way that history is consumed and understood around the world, providing access to the past for millions of users. Its mission is to enable users to engage with their past: 'Ancestry.com also operates a suite of online family history brands [...] all designed to empower people to discover, preserve and share their family history'[32] In moving to offer DNA services Ancestry opened up a huge new market. AncestryDNA, the brand created in 2012, took 5 years to become the biggest provide of Direct-to-Consumer Genetic Testing (DTCGT) worldwide, a position it has held ever since.[33] The market for DTCGT has ex-ploded in the past decade as around 250 firms offer tests directly to consumers.[34] This market is no longer health-focused, as it had originally been, and is now increasingly marketed around lifestyle choices. Around the world, customers are offered the opportunity to understand themselves in new genetic ways.[35]

Family history DNA testing has become a huge business. Ancestry was first made public in 2009, in an IPO raising $100million. It was then acquired by a private equity firm in 2012 for $1.6billion. 8 years later, a majority stake was bought by Blackstone Group Inc. for $4.7 billion.[36] These are enormous sums for what had begun as a small genealogical newsletter in the 1980s and had developed into a database-led family history resource.[37] What has driven this valuation is the size of the genetic database held by the group. By the time of writing, AncestryDNA has over 20 million customers in its database, dwarfing its number of subscribers for family history services (3 million). User numbers grew exponentially. In 2016, AncestryDNA had around 2 million customers; in February 2017 it had 3 million; in August 2017 5 million; in November 2017 6 million.[38] The figures have expanded every year due in part to clever marketing and advertising. By contrast its main competitor, 23andMe, has 12 million in its customer database. Ancestry's is the largest commercial DNA database in the world, but between these two companies there are around 32 million records held.[39] Even accounting for people who use both services this represents an enormous amount of genetic knowledge that has been built up in less than a decade. The family history companies that run these services, like MyHeritage, Ancestry, and FamilyTreeDNA, are evolving slowly into biotech companies as they work out ways to monetize and interrogate their new genetic data assets.[40] These companies have great influence on how genetic information is consumed and understood around the world, and furthermore a duty to contribute to education and awareness.[41] These companies are having a huge and unanticipated effect on El-Haj's 'genetic historical imagination', allowing millions of customers access to a genetically-informed and -inflected understanding of the past and its relation to them.[42] In particular, as suggested above, genetic data is being interpreted in terms of solid and quantifiable ethnic identity as much as it connects the user to multiple communities.

Vineet Mehra, executive vice president and chief marketing officer at Ancestry, argued that the combination of genealogy and genetics was aspirational and genuinely transformative:

> By making first genealogy and more recently genomics accessible and easy to use, we're committed to helping people unlock the past to inspire their future, because when you have a better sense of your own identity – it changes how you view the world and how you view your own future.[43]

This sense that genetic information can augment ordinary genealogy and contribute to a reconfiguring of both past and future is foundational the company's rhetoric. Through the intercession of better and more information, and particularly through the use of genetic data, individual identity is changed. The past becomes more understood, and that leads to shifts in the present. 'Unlock the family story in your DNA' runs Ancestry's marketing information, adding:

> Your DNA can reveal your ethnic mix and ancestors you never knew you
> had – places and people deep in your past where records can't always take
> you. Try AncestryDNA, and get a new view into what makes you uniquely
> you.[44]

Ancestry's marketing is keen to emphasise story and narrative, predicated up on
the idea of an origin: 'Your backstory is in your DNA: AncestryDNA can reveal
your heritage and connect you to family past and present'.[45] Similarly MyHeritage
advertises that 'The simple MyHeritage DNA test will reveal your unique ethnic
background, and match you with newfound relatives'.[46] FamilyTree DNA invites
you to 'Learn about your personal history and follow the path of your ancestors'.
These outlines illustrate key ways that genetic genealogy is sold. It can contribute
new ancestors and remake the story of 'you'. It contributes to your 'personal
history' and emphasises how 'uniquely you' you are. DNA-derived information
can enable the user to access a past that mainstream research – 'records' – cannot.
Central here is the word 'reveal', presenting the revelation of novelty, an intervention that could potentially change everything. Ancestry seeks to change the
way that we understand the past, enabling a new and aspirational sense of selfhood
to be constructed.

AncestryDNA and 23andMe have moved to realise the potential of their database by publishing scientific papers. AncestryDNA's work links the information
created by their family history users with genetic information generated by DNA
customers, including '322,683 pedigrees linked to genotyped samples in the
United States alone and over 20 million total pedigree annotations [that] allow us
to infer detailed historical portraits of the identified clusters'.[47] They are 'reading'
their two archives by linking the textual family trees with the genetic information,
allowing them to infer and analyse. However, because of the nature of the information their results are not replicable: 'we cannot make the genealogical and
genotype data widely available to the academic community in light of our commitment to our customers' (p. 9). This represents an interesting evolution of the
family history company, taking steps to read and interpret the archive that they
manage. Indeed scientific communities of all kinds – from population geneticists
to medical investigators – are interested to address the genealogical DNA archives.
The DTCGT company 23andMe, which gives users ancestry and family history
genetic information as well as health reports, recently announced a partnership
with the pharmaceutical company GlaxoSmithKline. Family history has become
interlinked with medical research, as the body's information becomes something
of contemporary use. It is important that we understand the ways in which the
DNA archive is being used commercially and how data generated about the past
might impact upon practices in the present.

Through the commercial generation of genetic information for family history,
companies are compiling major data archives. These collections of information provide huge amounts of information and revenue for such organisations, who are largely
private and often funded by hedge-fund and biotech venture capitalist investors.

As mentioned above AncestryDNA currently has over 20 million customer records in its database; other companies are much smaller. Evidently some family historians use multiple test services but this still leaves a large – and growing – amount of genetic data, tied to pedigree records. The ownership of this body of information, then, and the monetisation of it via commercial bio-banking and data-banking in an era of data medicine, is a very profitable enterprise.[48] The market for this information is growing fast, and expanding in unforeseen directions. Once sold on, the organisation, reading, and interrogation of such genetic ancestry information by researchers and medical practitioners might be considered as a kind of historical practice.

There is great potential opened up here, but also uncertainty about how this expanded information will change the way that family history works.[49] The massive expansion over the past couple of years in the databases held by the major family history companies raise compelling ethical questions: What should happen to DNA data now, and how will it be kept, curated, preserved and managed into the future? Where will this information be kept, and who will maintain it? Who owns the materials of the past, and access to those materials? The question of long-term storage and access is already becoming an issue as family historians who have had their DNA sequenced begin to die or take over management of the data of those who have passed away. Where should this information be held? Who should have access to it? It is not 'public' and not stored in any kind of public archive. Indeed, the creation of massive DNA databases can be seen to be part of a retrospective commodification of the materials of the past, making the dead work again by monetising their biological information. Certainly, the movement to opening up the DTCGT market has enabled the major companies to promote their product – genetic analysis – as equally important as the information generated by the state (census data, BMD data, education information). The companies commercialise access to the past through their subscription services. The development of the DNA archive, though, puts a value on individual genetic information. Customers become prosumers, volunteering their personal data and giving the organisation information which it then uses to create content.

It is also the case that millions of the DNA kits sold by the big companies are not used by practising family historians but are what might be termed leisure or throwaway usage, that is, bought for birthday or Christmas presents. They are activated, looked at, and then largely forgotten. Family historians reflect ruefully on the way that advertising of DNA kits seems to offer instant gratification, and many users drift away. This means that there is a large hinterland of information that is undeveloped, a large amount of inert data. The marketing push of the big companies has generated huge revenue and figures, but a lot of the information generated is relatively underused by the individuals involved. Concomitantly, the bigger the database, the better the result that can be generated, so a push to expand is important. Many people, therefore, are participating in the work of the company in generating results, even though they are not actively doing so. The companies need bigger databases to create useful and fine-grained results for their users. Each users' genetic information becomes part of the company's drive for expansion and profit, in a feedback loop. This means that customers who do not actively engage with the companies are still being 'used' by them, still generating information.[50]

The 'Amateur' Use of DNA for Family History

In January 2021 meezersqueezer, a self-confessed 'genealogy nerd' based in Philadelphia, posted a short video to the social media platform TikTok beginning 'Did anyone else take an Ancestry DNA test and uncover a family secret? Bc I did and didn't expect to'.[51] In the film they outlined how their family 'has always identified as 100% Irish', but after taking an Ancestry DNA test they were contacted by a Polynesian user named Fa'alongo. According to the company's database Fa'alongo had 'a high DNA match' with them. Having shared their family tree, meezersqueezer discovered that they and their match are related, and that their grandfather would seem to be the Polynesian user's half-brother. Their grandfather had throughout his life refused to answer any questions about his family history, saying 'the past is the past [...] so I never pressed him'. They also find out more: 'it didn't take me that long to find out my 3rd Great Grandfather on my dad's side was a confederate slave owner in Georgia'. They draw some conclusions from this:

> So it is really easy to sit here in present day Philly and feel really removed from slavery, it's a really big part of my family history on my Dad's side [...] and it's a really uncomfortable part that no one wants to talk about. But it's important for us to learn about it and acknowledge it

In the final posting, they reflect, 'So I really upset my family [...] but I was able to help out Fa'alongo and correct my own family tree'.

The original post by meezersqueezer was responded to hundreds of times and 'liked' over 43000 times, and subsequent parts of the same story were similarly popular. In addition to the comments in response, TikTok users 'stitched' with meezersqueezer, that is, used part of their clip as a way of introducing their own stories. The original post became a way of opening a discussion or prefacing particular revelations. The post generated hundreds of responses and 'stitches', creating debate and discussion from users around the world. Responding to the post and 'stitching' with it, user fretlessfeline, Cat McDonald, told a story to her then 108 followers about her own DNA test.[52] Her father gave her and 'the entire family' a DNA test for Christmas, in part to confirm that she was of Irish descent. Taking it, she discovered that, not only was she not Irish, but she had a precise paternity match with a user on the database who was not her father.[53] The actual match she had generated was an old family friend, a billionaire CEO, who refused to engage with her when she tried to connect with him. McDonald's story took off, sparking responses and shares, and was viewed by around a million other users.

It generated a wider debate about ethics, family, and paternity. McDonald now has 177000 followers and her posts have been liked 2.8million times. Her two posts on her DNA tests have been responded to over 27000 times. The story was also shared in the mainstream media, generating thousands of comments responding to articles on the websites of UK newspapers such as *The Daily Mirror*, *The Sun* and the *Daily Mail*.

We can use these two linked moments to more closely understand the phenomenon of 'leisure' DNA testing. They demonstrate the speed that social media can create discussion and bring together responses globally. They show the increasing usage of social media sites to share historical and genealogical stories and knowledge. The market for Ancestry and other DNA tests for family history is large and complex, and it has introduced genetically-derived knowledge about family history to a demographic way beyond that of mainstream genealogy. In particular, TikTok is generally considered to be a platform for younger users and is highly kinetic and fast-moving. It is not completely clear why Cat McDonald, of all the stories that were shared in response to meezersqueezer's original post, became so hugely popular. Possibly the sheer scale of the social difference was important, possibly McDonald's conspiratorial presentational style, whilst sipping wine, generated a feeling of connection. Many comments in mainstream media publications speculate on the veracity of McDonald's original story, which, whilst it reflects an online tendency for cynicism, does raise the possibility that the DNA reveal stories have become a kind of performative genre of their own. Certainly TikTok, similarly to other video- and image-based social media platforms, encourages performativity within its format.[54] The type of sociality that is created through the imitation and response aspect of TikTok films raises the 'Internet meme to the level of platform infrastructure', what Zulli and Zulli term '*imitation publics*'.[55] Certainly the importance of social media, and the genres that they encourage in their formats, should not be overlooked when thinking about the swift rise of genetic genealogy, and the continuing impact it has on historical sensibility around the world.

Genetic genealogy reveals an important new dimension in terms of how genetics allows new historical insight. DNA testing in family history allows for new connections with unknown ancestors, an opening up of knowledge and a complication of family history. Through the intercession of genetic data new emotional connections can be formed with ancestors. Genetic genealogy connects the user, through the company databases, to living people. Shared connections are created, allowing more finely textured knowledge of the genetic past. This combination of contemporary genetic self and historical investigation is central to genetic genealogy. However, it also introduces the possibility that family might be fractured, and forces users to consider the ethics of their use of genetics to understand their past. Whilst genetic information can provide new ways of understanding the past in relation to the present, it also raises profound ethical issues, and might also completely fracture that past. Cat McDonald used the experience to raise awareness of Non-Paternal Events (NPEs), asking her followers to donate to

DNAngels, a charity that works with people using DNA to identify their biological parents.[56] The transformative power of genetics in this instance, its revelatory impact, can challenge accepted family histories and overturn generations of secrecy.

In 2016 Marianne Sommer argued that users of DTCGT were at odds with wider scholarly critique of such services:

> While scholars from the humanities and social sciences have generated a critical discourse, warning about the potentially disruptive and essentializing effects of genetic identification, increasing numbers of the public partake in the markets as well as virtual worlds of genetic ancestry and build real-wordly connections on the 'history in their genes'.[57]

The exchange on TikTok, outlined above, demonstrate that this is to a certain extent true, insofar as the 'public' are creating huge markets for genetic information and finding new ways to connect and engage with each other and the past using this information.[58] At the same time, the accounts discussed above show a critical awareness of ethics and identity, and suggest new ways that genetic information has configured understanding of the way that the past relates to the contemporary moment.

For family history users, DNA data allows the verification of information, the breaking of 'brick walls', and the creation of further links that can be pursued.[59] Verification allows for a surety associated with information corresponding to existing information. It adds a level of confidence and solidity to the work, therefore. 'Brick walls' can be broken by the opening of routes for investigation and new information. This is a means of creating innovative and new ways of approaching the problem of the past. Finally, genetic genealogy allows users to investigate using connections made through genetic links. Users will take the information generated online by those who 'match' with them and add it to their work.

DNA generates large amounts of extra data for the family historian.[60] It is a highly useful tool when applied, that is, added to wider knowledge or understanding. Family historians are quite aware that DNA is not the final answer, and indeed their practice reflects an understanding that genetic genealogy is additional to, rather than replacing of, standard research. In this they demonstrate a way of being post-genomic that is largely practical. However, they also see genetic information as superadded to and distinct from mainstream, normative data that might be found in mainstream ways.[61] Given that much commercial testing is interested in presenting biogeographical ancestry results, the impact of genetic genealogy has also been to complicate models of the self in relation to race. There is a longstanding phenomenon, augmented by DNA testing, of contemporary peoples claiming different ethnic identities (see Chapter 3).[62] DNA information has accelerated this.[63] At the same time, genetic ancestry testing, whilst deeply problematic, has communicated a sense of complication and diversity to users. There is also some evidence that genetic testing might have very little effect on users' sense of their identity.[64]

What kind of evidence is this genetic data? Millions of amateur family historians see it as a new means to understand their past. It has to be 'read' and understood, and increasingly family history groups are generating tools for displaying and visualising DNA information. Therefore DNA becomes a new form of historical information, used alongside archival investigation or research in local history libraries. It has the status of 'evidence', contributing to historical understanding and. awareness. It is, though, in addition to those archives and local history libraries. Beforehand, family history evidence was gathered through engaging with (mainly) public institutions that had collected it for particular reasons. This means that the type of information gathered was precise. Family history is largely drawn from data regarding birth, marriage, death, and occupation (that is, information gathered generally by censuses and churches in Western countries). DNA data is a different kind of evidence, mostly gathered and stored by large commercial organisations. Individuals may have also downloaded and worked on their information. Hence, research in this area, particularly in the future, will entail a clear understanding of the purpose of collection and an awareness, particularly, of the commercial drive of such databases.[65] There is now a huge amount of new genetic historical data in existence, but it is not easily accessible or 'readable' and historians will need to develop tools and methodologies in this area.

Therefore, DNA data has the ability to change significantly the way that the family history practitioner approaches the past, from adding solidity to their findings to opening new spaces for research. They can generate entire datasets that had not been available to them and pursue new directions. Individual users can download their information and analyse it themselves, should they wish to; or they can use the tools provided by the commercial companies. There are also increasing numbers of open-source software packages and user-generated tools to manipulate, visualise, and analyse genetic information. This user-generated approach means that the family history community is at the forefront of developing techniques for understanding genetic genealogical data. Training and knowledge is relatively sparse, and for the most part the family historians using DNA are self-taught. This leads to a wide disparity in knowledge and practice. That said, family historians are rapidly pooling their knowledge through societies and online groups. Collaboration is very important, and this has led to the development of crowd-sourced tools and databases (discussed below). Family historians tend to be highly self-conscious about their practice.[66] Therefore they increasingly have developed methods to approach DNA information. Indeed, some family history practitioners are now highly skilled in DNA analysis and this has led to their being used by businesses working on cold cases (see Chapter 4).

The new information generated from DNA can augment emotional relationships with the past. This affective aspect of the engagement with the past is important. Genetic genealogy is predicated upon the bodily connection of the individual user to their past. Genetics allows them to seemingly reach back and create new connections with their history. These connections are fundamentally important to their sense of self. They are also emotional, allowing a reconfiguring of the contemporary participant through the intercession of scientific data. There is often a poignancy or melancholy about family history, and DNA evidence is now part of that aspect. The reorganisation of investigative practice by family historians therefore leads to a reconfiguring of the idea of the family and of the self.

DNA data promises new connections and to open up the practice of family history. It can extend the scope and compass of research, taking practitioners to different areas geographically and chronologically. It allows the investigation of female lines, in particular, which can be transformative given that records for women are far less thorough than for men, if they exist at all.[67] DNA also provides seemingly incontrovertible proof, revealing 'truths' that might have been lost in the archive, never recorded, or hidden in the documentary evidence. This leads to a new assurance about historical investigation, a sense of finality. The promise of DNA research is that it will be bulletproof, 'solving' the past in a way hitherto impossible. For particular groups such as African Americans or Jewish communities, DNA gives a way of challenging historical abuses and reconnecting with families once thought lost. Genetic genealogy can seem solid and authoritative, and indeed this is exactly how it is marketed.

Genetic genealogy shifts the way that family history narratives work, in particular, and hence changes the consideration of selfhood that is at the centre of such practice.[68] DNA can change the concept of what a family might be, how it might consist, and how it is configured. However, it may also bring into the light long-hidden secrets and problematic information, and this leads to new ethical considerations for those who practise it. With the massive expansion of the genetic genealogy databases, the incidence of people discovering non-paternity events in their background, particularly, has increased hugely.[69] Customers have discovered new siblings, entire new families, and proof of behaviour that challenges their idea of family members. Genetic genealogy is a highly efficient way to search for missing relatives. Users have discovered revelations about relations, discovering aspects of shame and guilt thought long-buried. This has allowed a certain amount of historical reckoning to happen, particularly amongst white communities in settler colonial countries such as the USA, Canada or Australia. The data provides unexpected and often unwanted change to the structures of families. At the same time, customers have 'lost' family, suddenly finding out that what they thought was the case is wrong.[70] This challenges the idea of the family, the structure of the community even. Through investigating the past using DNA, users can reconfigure their present in ways they had not anticipated. DNA becomes a light shining on uncomfortable truths. This has ramifications often in the contemporary world, as genetic genealogists discover things about themselves or their families that had been kept quiet, ignored, or not known. Therefore, the intercession of DNA is not always welcome.

Family historians are increasingly worried about the ownership of data and their individual genetic privacy. Bioprivacy is increasingly an important issue for family historians. Genetic data does not simply implicate the individual, it includes family members from parents to children to siblings, even implicating first, second and third cousins. Therefore, there are issues of privacy, consent, and ethical duty to consider when approaching this type of investigation, issues that are often not to the forefront of the commercial drive to expansion of the database. DNA also 'lives' beyond the user's lifespan. Many users maintain the profiles of people who have already died, meaning that genetic information can be used by other researchers after the death of the original participant. This continuation means that genetic data has a spectral aspect. This is the case, too, for the ways in which genetic information allows an engagement with the melancholic ghosts of the archive; through new data, those long-since dead can be reanimated and given new meaning. Such playing with the dead through genetics is relatively common. One of the community-based websites, GEDMatch, has a tool called Lazarus which allows the user to 'recreate' the DNA of their dead relatives. Through elaborate calculation the tool can speculate about the genetic make-up of dead people.

Genetic genealogy has the potential to disrupt or challenge normative family structures. Such approaches can reveal new ways of thinking about the past, of understanding the contemporary moment, and of relating to wider communities. The massive market for DTCGT demonstrates an appetite for such knowledge, but the consequences of this huge expansion in information are yet to be fully understood. There are new archives, new ways of understanding the human, and new connections being made daily. DNA-derived information about the past changes the notion of self in the present and undermines standard historical structures of knowledge. It challenges the centrality of particular types of evidence.

Genetic Genealogy Documentaries

We can see the impact of DNA on individual understanding of the past when considering its manifestation in popular culture. In particular looking at historical documentaries which utilise DNA enable us to see how concepts of the family and of the self in relation to the past have been reconfigured in the past decade. DNA evidence is enabling a new way of knowing the past, either communal or individual, and these television shows demonstrate the widespread ways that this new knowledge is being transmitted and becoming normalised.

There are numerous shows about genealogy on television, all to some extent developments of the highly successful series *Who Do You Think You Are?* (*WDYTYA?*, BBC, 2004-), *Long Lost Family* (ITV, 2011-), and the series in the US fronted by Henry Louis Gates, Jr. (see discussion below).[71] *WDYTYA?* has run continually for 17 series and has been sold around the world. The format has been successfully exported to the USA (10 series, 2010-) and 18 other countries; *Long Lost Family* has similarly had global success, and heavily promotes AncestryDNA in the USA.[72] The series has led to several further documentary series about family history including *Every Family has a Secret* (SBS, Australia, 2019) and *Secrets of the Workhouse* (ITV, UK, 2013).

The model of *WDYTYA?* is celebrity-driven, as each episode sees a famous figure investigate their family history. The series is predicated on revelation and mobility, with individuals moving around the country and the world to look for their roots.[73] Genetic genealogy is very rarely used in *WDYTYA?* Celebrities meet experts and visit important locations, using documentation and databases to track down unknown parts of their family history. These shows often struggle to impose order upon the history that is unearthed: 'genealogical documents often raise more questions than they answer, since inconsistencies, surprises, and scandals are at the core of family history research'.[74] There is an emphasis on shots relating to travel, from trains to car journeys, emphasising mobility and novelty as well as the motif of investigation transposed into physical movement.

Long Lost Family involves members of the public being reunited with relatives and similarly emphasises movement around the world. The show began to use DNA evidence in 2017 and it is now a staple of the show; it is pioneering as a popular TV series in using genetic data to investigate ordinary families, although there is less evidence on the historical investigation and more on the process: 'we find people nobody else could trace, We uncover incredible family secrets'.[75] The two new shows screening in 2021 demonstrate the increasing importance of genetics in driving genealogy, and illustrate the evolution of the genealogical documentary. They allow us to see how genetics has become mainstream in interpreting and presenting family history, and in particular to illustrate how the key work undertaken by DNA is revelatory. As Claire Lynch has argued, *WDYTYA?* and *Long Lost Family* represent 'biogravision', a combination of social history documentary with personal and celebrity revelations influenced by reality-TV.[76]

The fact that DNA family history has become mainstream as a type of approach to the past is illustrated by two prime-time television shows focusing on genetic genealogy which both began on prime-time television in the UK in 2021: *DNA Journey* and *DNA Family Secrets*. The two series demonstrate how 'family history' is increasingly being seen as driven by DNA revelations. They show us the evolution of the genealogy documentary, and indeed the historical documentary, through their focus on the body, on emotion, and on affective connection between then and now. They also show the increasing power of the genealogical companies and their influence on documentary production.[77]

The pilot of *DNA Journey* screened in 2019, before the format was developed to series length. The pilot episode featured British TV presenters Ant & Dec, who discovered it likely they share a very distant relation. In 2021, after this initial episode, ITV formally launched a series of *DNA Journey*, their intervention into the evolving mode of the genealogical documentary. Reviewers recognised that this was an attempt at revising 'an otherwise tired format', claiming it to be 'ITV's stab at picking off *Who Do You Think You Are?*'[78] Reviewers also praised the 'lighter touch' of the series in comparison with the 'traditional *WDYTYA?* solemnity and lachrymosity'.[79] The show is confidently light and sensational, with a pop soundtrack including 'Family Tree' by the Kings of Leon. In style, the show is attempting very much to present a challenge to different types of genealogical documentary – particularly *WDYTYA?* – that are in comparison relatively dry and historical. This series is interested in presenting an alternative type of investigation, more irreverent but also emotionally acute. The editing, particularly of the opening sequences – drone shots of landscape and cities, split-screen, tight rocky soundtrack, freeze cuts, clips of the main participants laughing hard at each other – echoes the kind of caper film making of Steven Soderbergh's 2001 *Ocean's Eleven* in order to emphasise that this is 'modern', new, and dynamic. The series pairs celebrities, encouraging banter and competition as well as enabling a certain reflection. In particular, building on the first pilot show which focused on famous friends/presenters Ant & Dec, the show emphasises friendship and companionship. The participants undertake the 'journey' together: 'family is so key to us both and that is one of the reasons that we wanted to do this journey'.[80]

Compared to *WDYTYA? DNA Journey* is particularly focussed on English nationhood and the focus is predominantly white, male, and relatively heteronormative (of the 8 white participants there is one woman and one gay man; the majority of ancestors focused on in the show are male). The emphasis is on the relationship between family and self-identity. In each show the initial focus is on the space of England as is often the case with genealogical programmes. There are several key establishing shots of green landscape and coast. Mobility, always a motif in genealogy shows, is underlined by shots of cars and taxis. *DNA Journey* has a focus on literal journey-making and movement, with shots of cars on the motorway, and repeated scenes shot inside cars as the participants are driven around the country. Mobility is one of the ways that genealogical series formally introduce ideas of connection and migration, of moving outside of the individual's normal experience to something more interconnected. In genealogical documentary this initial focus on Englishness is often diluted by the mobility of the family history, part of the revelatory project being to demonstrate that the foundations of the modern nation are more complex than hitherto considered. However, the movement in this series tends to be largely within the UK, and in particular within England. This is emphasised by the participants' interest in their fixed identity and concern that it might be challenged. At the end of the first episode Freddie Flintoff's DNA breakdown is revealed, a 'map of your DNA' with an ethnicity estimate, with the comment 'You could not be any more Prestonian [from Preston] if you tried'. Flintoff is 'so happy' 'Knowing that I'm from the place I love and I feel most comfortable and at home with'. For Flintoff genetic genealogy substantiates a sense of particularly connection and identity, although his concern that it might *not* do shows the revelatory revisionist potential associated with DNA in TV programmes like this.

The (national) mobility of the show reflects its engagement with the past through genetics: in *DNA Journey* the emphasis is on a movement from ignorance to knowledge. This really is an emphasis on self-knowledge, though, a key trope for genealogical documentary. Flintoff opens by saying 'what people see on telly [...] is not necessarily me', and the suggestion by his friend Jamie Redknapp is that 'there might be some of the things we find out in our DNA that that makes us, or gives us the traits that we have'.[81] Flintoff worries that his sense of identity might be challenged: 'I've got this idea that I'm very English, I come from Preston and I'm from the North [...] I don't wanna find out I'm from a long line of bad people'. However, he reflects that 'It might explain a few things'. As mentioned above, Flintoff's concern with his identity is entwined with an investigation which seems to secure models of lasting and historic 'Englishness'. He is also looking, in his rhetoric at least, for some kind of answer or explanation for his contemporary selfhood.

The DNA intervention is very low-key, with participants receiving alerts on their phones and responding to briefings in the street by Brad Argent, an expert from Ancestry. Information arrives, no explanation, simply assertions of connection and actually quite a lot of emphasis on documented family trees than DNA work. Indeed, the lack of sophistication is conscious, as explained by Argent: 'Our history is more accessible than ever before. The vast bulk of the research for #DNAJourney was done with a laptop and a phone – proving that you don't need privileged access to uncover great stories'.[82] This challenge to 'privileged access' suggests a route for the genealogy documentary, a move away from dramatizing research towards a concern with experience.

Made for the BBC, *DNA Family Secrets* is less focused on motifs of movement and more, as the title might suggest, on revealing what is not known. Introduced by Stacey Dooley, a well-known journalist who makes authored documentaries about social issues, the show seeks to 'unlock life-changing secrets through the power of a DNA test'.[83] Dooley's voiceover begins each episode: 'Many of us have questions about who we really are [...] and what our future holds [...] Now amazing advances in DNA technology mean those questions can be answered for the very first time'. The focus, therefore, is on what is not known, what can be revealed by the intervention of new technology: 'we're going to help people unlock life-changing secrets through the power of a DNA test'.[84] The link is direct between the novelty and progress represented by DNA work, and the secrets held by families. DNA here acts as a modernising technology, giving participants access to a past that had been denied them. It is also something strong, a 'power' that can realign the past and reconfigure the present. 'Our DNA doesn't lie', says Dooley in voiceover, conferring absolute credence on the science and emphasising its precision, accuracy and revelatory force: 'but are we ready for the truth?'

For a documentary about DNA and titled *DNA Family Secrets*, there is little overt science, similarly to *DNA Journey*. This lack of laboratory (or library) work is the new direction in genealogical documentary, and this has implications for the understanding of DNA work in relation to family history, as well as the ways that genetic genealogy might contribute to the genetic historical imagination. Genetic information has become increasingly powerful, but also divorced from scientific labour particularly. Its rhetoric is in this context that of connection and revelation. For the two documentaries, whilst DNA gives them their purpose and to a certain extent their credence, the concern is with the impact of genetic information upon the individual (celebrity, participant) rather than the workings of the evidence itself. Turi King, the expert (well known for her work on Richard III, see Chapter 1), appears in repeated segments with each participant, but these take place in a comfortable room on easy chairs. There is one brief shot of her in the lab. King draws family tree diagrams, and that is the extent of the medical aspects of the programme. This is a conscious strategy to avoid particular tropes of science film (white coats, laboratories) and instead ensuring that the genetic work seems to be incorporated into the life of the show.[85] King is throughout presented as an approachable presence rather than distanced and laboratory-based. She explains the science relatively quickly and uses a language herself of revelation and authenticity: 'The thing about DNA is that it can help you uncover the truth. Are you ready for that?' The lack of any real account of what genetic work is being done, outside of swift and often analogous explication by King, renders the analysis hidden and unconsidered. Even the action of taking a swab is rendered quite low-key, as participant Bill says 'that's a little bit of me' about the saliva collected on a swab. Another participant, Margaret, is filmed carefully spitting into a tube in front of King. There is a clear distinction made between 'expert' and everyone else, but this is more in the structure of the show. The protagonists ask King their 'DNA Question' and she explains how she is going to work. Her introduction of DNA technicalities is consciously engaging. When explaining centimorgans, for instance, she explains that this is a 'fancy term for a unit of genetic measurement – how people share sections of their DNA'. Very little is given in terms of hard data or information, and indeed the 'work' of genetics is not huge, given that DNA forms the basis of the programme's novelty. This is important in making the DNA work non-threatening and part of normal life, but it also has the effect of placing the genetic analysis into the background of the programme.

The involvement of Dooley ensures that the focus is on the emotional journey and the individual experience. Her important contribution to the show is empathy when speaking with those who are discovering important things about their lives, and hence she helps situate the show as being primarily focused on revelation and emotion. In *DNA Family Secrets* the focus on movement becomes formal, as each section is introduced by drone shots of motorways and railways (similarly to *DNA Journey*). Dooley's voiceover is set against shots of streets, houses, schools, before moving in quickly to the particular story that is going to be worked through. This produces an effect of shifting from wide focus to narrow and works neatly to suggest the picking out of individual narratives from the wider history. Key to the way the programme presents itself is the idea of *knowledge*, as DNA allows participants to 'unravel their family secrets'. When asked about their involvement participants articulate this drive to understand: 'I'd just like to know where I come from'; 'I'd like to know before it's too late'; 'I'd like to know where I'm from, really [...] I'd like to know who my Dad is'. This focus on knowledge is part of the show's structure, as each participant is asked 'what would you like to know' using their DNA. Hence the show dramatizes the idea of DNA as explanatory and revelatory, as filling in gaps in knowledge, and exposing secrets. 'That's quite a revelation' says Bill, one of the participants. This language is couched in a discourse of veracity and authentication, with knowledge (conferred by DNA) leading to revelation and a new 'truth'.

Bill also adds a note of melancholia to the programme, saying when he meets cousins for the first time 'such a lot of life that has gone by'. Richard, another protagonist, wonders about the man who brought him up: 'all my life has been like a play, all the fond memories I have of him its just like, He didn't know, I wonder if he knew?' Yet the series generally does not have time for this type of sadness and absence, preferring to be more focussed on the positive aspects of reuniting families. The link between historical revelation, heightened genetic awareness, and emotional wholeness is continually made. *Knowing* becomes something possible through genetic genealogy and is linked to an understanding of the self in the contemporary moment; the past is changed and the present is enlightened. In these shows historians are either marginalised or not even included, yet the documentaries are making strong claims for the ways in which we might investigate, know, and understand the past. Historical understanding, as rendered and communicated through documentary, is being reconfigured through the use of a new type of evidence.

Reviews of the series focused on the emotional work that it was doing: 'to not be moved watching relatives meeting for the first time, you'd have to be very cynical indeed' wrote Rachael Sigee.[86] The reception of the series was generally interested in this aspect, emphasising how moving the programmes were. Writing in *The Mirror*, Sara Wallis noted the shifts in the representation of genetic science that the show demonstrated:

> Remember when the greatest TV tension was watching Jeremy Kyle reveal the results of a DNA test to warring couples? In those days the tests only passed through the hands of medical professionals in white coats or the all-hallowed Kyle research team. We've come a long way since then [...] In true millennial style, Stacey and co just have to tap a few buttons to access a massive DNA database thanks to some 25 million people worldwide doing a home test.[87]

The difference in the ways that DNA is now presented on television gives Wallis some good material with which to note how things are done 'these days'. For her the ease with which Dooley's team can address the DNA archive is generational ('true millennial style') and suggests a shift from more challenging genetic work to contemporary ease of access. Wallis's irony reminds us of the way that DNA had generally been considered in popular culture – in daytime TV shows using paternity tests – and notes the shift towards using DNA as a restorative tool for families. Both types of show depend on revelation and use genetics as a means of assurance. Wallis also points out that much of the work that is being undertaken in *DNA Family Secrets* is being done by a silent contribution from those users of commercial databases who have uploaded their results. This is referred to in the series, as we are repeatedly told that 25 million people have uploaded their details to the worldwide 'network of DNA databases'. The new ease with which the 'millennial' might know their genetic make-up is dependent on an archive created over the past decade. Wallis's points about smoothness reflect the way that *DNA Family Secrets* presents the genetic science. It is something that can be easily investigated, with clear and comprehendible results. Finally, Wallis's judgement on the show, like Sigee, depends on how it provokes an emotional response; she cries three times watching the 'emotional' and 'gripping' show.

The focus on interiority and the understanding of self is common in genealogical celebrity shows.[88] The concern with self reflects the emphasis on individual becoming in contemporary reality-influenced documentary.[89] Indeed this demonstrates the evolution of the documentary form over the past decade. It is also an interpretation of family history which suggests a teleological, chronological movement towards the contemporary moment. The individual participant uses genetic genealogy in seeking to explain themselves, considering that the genetic past contributes strongly to their contemporary identity, without them knowing. The journey here is internal, exposing something internal but hitherto unacknowledged. In these shows historians are either marginalised or not included, yet the documentaries are making strong claims for the ways in which we might investigate, know, and understand the past through the intercession of genetics. Alongside 'traditional' methods and historical sites – locations, objects, documents – genetic information is bringing the past into a different focus. Historical understanding, as rendered and communicated through documentary, is being reconfigured through the use of this new type of evidence.

Genetics Driving Historiography: African American Genealogy

Henry Louis Gates, Jr's highly successful *African American Lives* (2 series, PBS 2006-), *Face of America* (PBS, 2010) and *Finding Your Roots* (7 series, PBS 2012-) shifted fundamentally the way that genealogy was perceived in the United States and also contributed to a new way of thinking about African American identity. In the series and accompanying books Gates, Jr. uses family history and genetic genealogy to establish revisionist historiography. The first episode of *African American Lives* itself includes a brief discussion of *Roots*, Alex Haley's highly influential television series and book from 1976-7.[90] In an opening episode Gates, Jr. discusses *Roots* with Quincy Jones, who had produced its music. *Roots* had presented Haley's narrative of tracing his past to ancestors in Gambia. Writing of the series in 1977 in *The Black Scholar* Chuck Stone argued that it gave 'many blacks who had not known who they were [...] an ennobling sense of their pastness'.[91] The discussion of *Roots*, what Louis Gates, Jr. terms 'that first journey back to Africa', establishes a direct connection between the impact of that series and the profound effect *African American Lives* will have on African American communities. Yet, as is noted by Gates, Jr. in this opening episode, *African American Lives* has the ability to go beyond *Roots*.

This sense that 'modern' techniques for investigating, in particular the use of genetic information, can be transformative for the individual and the community is key to the epistemological and historiographical shift that Gates, Jr's work has effected.

African American Lives used DNA genealogical techniques to allow participants to reach to a history that had been denied them: 'when the paper trail would end, as it inevitably did, in the horrid darkness of slavery, we traced our African roots through our DNA' (p. 11). In Gates, Jr's work, genealogy is reclamatory and politically revisionist (see discussion in chapter X). As Ongaga describes his practice, 'doing family trees adds specificity to the raw data from which historians can generalize about the complexity of the American experience'.[92] Genealogy is deeply important in understanding community and historical experience, but it also enables a challenge to narratives of nation. *African American Lives* begins with Gates, Jr. at Ellis Island (with establishing shots of the neighboring Statue of Liberty), traditionally and symbolically the location of a particular model of American immigrant identity and national understanding. Gates, Jr., though, immediately challenges the centrality of the Ellis Island immigrant experience:

> For millions of immigrants a gateway to the new world. Today their descendants come to this place seeking a connection to the past. I envy my friends who can come here and celebrate their ancestor's journey and trace them through the records so diligently compiled here [...] Unfortunately there is no Ellis Island for those of us who are descendants of the African slave trade. Our ancestors were brought to this country against their will.[93]

In opening at Ellis Island, the series posits an alternative set of narratives of American history and identity. Implicit is a criticism of the archival state, 'diligently' keeping records to establish a particular nation but denying this to slaves and their ancestors. This lack of archive therefore marginalised peoples and cut them from history entirely; through DNA, they are able to take 'the once unimaginable journey into the Black past'.[94] The word 'unimaginable' is doing a lot of work here. On the one hand, it points towards the revelation that the series is predicated upon, the opening up of new research and new ways of understanding. On the other, Gates Jr's words clearly assert the fact that history, for African Americans, was something they were not allowed to imagine, dream of, consider their right. Key in enabling this new knowledge and this challenge to racist structures and infrastructures of memory is DNA:

> For generations we have been unable to learn about our African heritage or our family trees. But what if we could trace our roots? What stories would we discover, what ancestors would we meet? What if we could even travel through time, cross the Atlantic Ocean and find where our ancestors were from? Now, thanks to miraculous breakthroughs in genealogy and genetics, we can begin to do just that.[95]

The 'connection to the past', actively denied African Americans is now decisively enabled, through the radical intervention of DNA. Genetics provides evidence that can bypass the racist archive and the acts of erasure and forgetting imposed upon African Americans for centuries. *African American Lives* followed 'nine remarkable African Americans' as 'they begin to find their place in history'.[96] The focus on celebrity was important, suggesting that these figures represented modern America, and therefore that their history should be considered significant in the development of the country. The series calls their experience 'one epic journey', conflating the nine together and suggesting a teleology towards the present. The opening credit sequence links each participant's name, implying a connection between them all; that connection being race, but also a sense that they – and their wider community – has been historically considered as one people, and historically marginalised as a consequence.

Each of these series uses DNA, blending it with traditional techniques of archival research: 'to uncover their roots we've used every tool available to us. Genealogists helped stitch together the past from the paper trail their ancestors left behind … while DNA experts utilised the latest advances in genetic analysis to reveal secrets hundreds of years old'.[97] The science is constantly presented as new, transformative, groundbreaking: 'Revolutionary advances in DNA analysis will allow all of my guests to get their first glimpse of their African heritage'.[98] Each episode is highly static, shot as an interview between Gates, Jr. and the subject of the programme who works through their 'book of life'. Intercut are cutaway shots of historical documentation, brief interviews with experts, and some location shots; the focus, though, is on the participant and their engagement and recollection. DNA is used to 'uncover connections' not seen and reveal the 'unexpected places that their ancestors once called home'.[99] *African American Lives* used genetic tests to reveal hitherto unthought of ethnicity, particularly the percentage of 'European' DNA they might have and connections to shared ancestors: 'Are you surprised to learn that you have that much European ancestry?' he asks the actor Don Cheadle.[100] Indeed, whilst we should be careful of 'these texts' conflation of DNA genealogy and race', it is often the case that Gates, Jr. uses DNA as a means for presenting complication and historical nuance to individual historicity. The shows work to demonstrate that genetic evidence can decisively complicate the way that individuals think about their past, their nation, and their identity in the present; but it can also challenge a way of remembering, bringing new and alternative evidence that allows a new narrative to be articulated. For these shows, DNA evidence is both new and old simultaneously. It is revelatory, but the action is to bring into the light what was always there. It is therefore comforting and radical at the same time, exposing racist structures of public memory whilst suggesting a nation reborn.

Henry Louis Gates, Jr's arguments about the effect of African American genetic genealogy on historiography and historical documentary practice can be seen enacted in the major series *Enslaved* (BBC/CBC, 2020-). *Enslaved* utilises a similarly genealogical approach, connecting the celebrity in the contemporary moment with a wider history, and, crucially, deploys genetic connections to provide new insights into the experiences of the African diaspora. It is a revisionist series, insofar as it reconfigures what is known using DNA and genealogical investigation as a tool. For this programme the actor Samuel L. Jackson investigates the slave trade, 'bringing to light our […] forgotten history' to educate the public: 'still millions of descendants don't know where their ancestors actually came from'. The work of DNA is brief but highly significant, enabling the show to exist, acting as a conduit between then and now, and situating Jackson. In an opening sequence he visits a cousin to talk about their family history, and remarks 'we're able to do something uncommon for most African Americans […] trace our lineage back to the days of lineage'. This 'word of mouth' family tree can only get him so far, to the absent historical violence of slavery, and then next step is the DNA test. *Enslaved*'s first episode is based on Jackson's experience of Gabon and his travel around the country, visiting key sites and uncovering the evidence of the transatlantic trade in slaves. The focus on Gabon allows the documentary to begin formally with origin, linking Jackson as the contemporary commentator with the historical moment. This would not have been possible without the genetic link which acts as a direct enabler of the connection. Evidently the connection between Jackson and Gabon is there (if hitherto not known), but it lacks clarity, sharpness, and visibility without the DNA test. Jackson's case embodies the revisionist historiography that Gates, Jr., argues for. Without the DNA Jackson's family history knowledge is limited; with it, he can suddenly connect to a new country, a new set of communities, and a new sense of himself. Furthermore, the intervention of the genetics provides the mainspring for the documentary, allowing it to happen. DNA information here opens up ways of knowing, narrating, and experiencing the past, both individually and collectively.

Notes

1 Tanya Evans, 'Secrets and Lies: the Radical Potential of Family History', *History Workshop Journal* 71, 2011, pp. 50–73.

2 Ashley Barnwell, 'Convict shame to convict chic: Intergenerational memory and family histories', *Memory Studies* 12:4 (2017), 398–411 and Wendy Bottero, 'Practising family history: 'identity' as a category of social practice', *British Journal of Sociology* 66:3 (2015), 534–6. See also Alison Light, *Common People* (Harmondsworth: Penguin, 2014).

3 See the essays collected in the special issue of *International Public History* 2:2 (2020), https://www.degruyter.com/journal/key/IPH/2/2/html.

4 Laura King and Gary Rivett, 'Engaging People in Making History: Impact, Public Engagement and the World Beyond the Campus', *History Workshop Journal,* 2015 80:1, 218–233, p. 229.

5 See Jerome de Groot, "On Genealogy," *The Public Historian* 37:3 (2015), 101–27.

6 On the Church and its family history activities see Julia Creet, *The Genealogical Sublime* (Amherst, MA: University of Massachusetts Press, 2019).

7 Jerome de Groot, 'The Genealogy Boom: Inheritance, Family History, and the Popular Historical Imagination' in *The Impact of History? Histories at the Beginning of the 21st Century*, edited by Bertrand Taithe and Pedro Ramos Pinto (London and New York: Routledge, 2015), pp. 21–34.

8 Information on the company can be found on their corporate facts website, https://www.ancestry.com/corporate/about-ancestry/company-facts [accessed 23 July 2021].

9 Richard Tutton, '"They want to know where they came from": population genetics, identity, and family genealogy', *New Genetics and Society* 23:1 (2003), 105–20.

10 Gísli Pálsson, "The Web of Kin: An Online Genealogical Machine," in *Kinship and Beyond: The Genealogical Model Reconsidered*, ed. Sandra Bamford and James Leach, (New York and Oxford: Berghahn Books, 2009), pp. 84–110.

11 Mark A. Jobling, 'In the name of the father: surnames and genetics', *Trends in Genetics* 17:6 (2001), 353–57.

12 George Redmonds, Turi King, and David Hey, *Surnames, DNA, and Family History* (Oxford: Oxford University Press, 2011).

13 *Surnames, DNA, and Family History*, p. 208, 209.

14 Bryan Sykes, *The Seven Daughters of Eve* (London: Bantam Press, 2001).

15 See Colin Renfrew, *Prehistory: the Making of the Human Mind* (New York, NY: Random House, 2008) and Yuval Noah Harari, *Sapiens: A Brief History of Humankind* (New York, NY: Harper Collins, 2015).

16 Abu El-Haj, p. 144.

17 Nicole Bradford, 'Riding the Genetic Revolution', *Houston Business Journal*, 24 February 2008 https://www.bizjournals.com/houston/stories/2008/02/25/smallb1.html?page=all [accessed 4 March 2021].

18 https://www.familytreedna.com/ [accessed 4 March 2021].

19 Abu El-Haj, pp. 144–51.

20 See Sarah Abel, 'Of African Descent? Blackness and the Concept of Origins in Cultural Perspective', *Genealogy* 2:1 (2018), https://doi.org/10.3390/genealogy2010011.

21 Alondra Nelson, 'Bio Science: Genetic Genealogy Testing and the Pursuit of African Ancestry', *Social Studies of Science* 38 (2008), 759–83 (761–2).

22 See the essays in the special issue 'A Market in the Making: the Past, Present and Future of Direct-to-Consumer Genomics', *New Genetics and Society* 36:3 (2017).

23 P. Borry, M.C. Cornel and H.C. Howard, Where are you going, where have you been: a recent history of the direct-to-consumer genetic testing market', *Journal of Community Genetics* 1:3 (2010), 101–106.

24 'The Gift of Spit', p. 240–2.

25 'The Gift of Spit', p. 237.

26 Stuart Hogarth, Gail Javitt, and David Melzer, 'The Current Landscape for Direct-to Consumer Genetic Testing: Legal, Ethical, and Policy Issues', *Annual Review of Genomics and Human Genetics* 9 (2008), 161–2 (162).

27 Heidi C. Howard, Pascal Borry, Bartha Maria Knoppers, 'Blurring Lines', EMBO Reports 11 (2010), 579–82.

28 Anna Middleton et al, 'Direct-to-consumer genetic testing: where and how does genetic counselling fit?', *Personalized Medicine* 14:3 (2017), https://doi.org/10.2217/pme-2017-0001.

29 See the discussion of the ethical implications of this phenomenon in Anne Harris, Sally Wyatt and Susan E. Kelly, 'The Gift of Spit (and the obligation to return it)', *Information, Communication & Society* 16:2 (2013), 236–57 and Stuart Hogarth and Paula Saukko, 'A market in the making: the past, present and future of direct-to-consumer genomics', *New Genetics and Society* 36:3 (2017), 197–208.

30 See Libby Copeland, *The Lost Family: How DNA testing is upending who we are* (New York, NY: Abrams Press, 2020).

31 Adam L. Horowitz et al, 'Consumer (dis)interest in genetic ancestry testing', *New Genetics and Society* 38:2 (2019), 165–94.

32 http://www.ancestry.co.uk/cs/legal/Overview [accessed 15 April 2021].

33 See Jerome de Groot, 'On Genealogy', *The Public Historian*, 37:3 (2015), 102–27.

34 A. M. Philips, 'Only a Click Away – DTC Genetics for Ancestry, Health, Love … and More: A View of the Business and Regulatory Landscape', *Applied & Translational Genomics* 8 (2016), 16–22.

35 See Anne Harris, Sally Wyatt and Susan E. Kelly, 'The Gift of Spit (and the obligation to return it)', *Information, Communication & Society* 16:2 (2013), 236–57.

36 Heather Perlberg, 'Blackstone Reaches $4.7 Billion Deal to buy Ancestry.com', *Bloomberg*, 5 August 2020, https://www.unilad.co.uk/news/ancestrys-database-of-peoples-dna-was-just-bought-for-nearly-5-billion/ [accessed 23 Feb 2021].

37 See Jerome de Groot, 'The Genealogy Boom: Inheritance, Family History, and the Popular Historical Imagination' in *The Impact of History? Histories at the Beginning of the 21ˢᵗ Century* ed. Bertrand Taithe and Pedro Ramos Pinto (London and New York: Routledge, 2015), pp. 21–34.

38 https://www.ancestry.com/corporate/newsroom/press-releases/ancestry-surpasses-5-million-people-dna-database-giving-customers-even-more; http://www.globenewswire.com/news-release/2017/11/28/1207093/0/en/UPDATE-AncestryDNA-Breaks-Holiday-Sales-Record-for-Black-Friday-to-Cyber-Monday-More-Than-Triples-Kits-Sold-Versus-2016.html

39 Christine Farr, 'Consumer DNA testing has hit a lull – here's how it could capture the next wave of users', *CNBC.com*, 25 August, 2019, https://www.cnbc.com/2019/08/25/dna-tests-from-companies-like-23andme-ancestry-see-sales-slowdown.html [Accessed 3rd June, 2020].

40 de Groot, 'On Genealogy', p. 12., and *passim*.

41 Belen Hurle et al, "What does it mean to be Genomically Literate?," *Genetics in Medicine* 15:8 (2013), 653.

42 Christine Scodari, 'When Markers Meet Marketing: Ethnicity, Race, Hybridity, and Kinship in Genetic Genealogy Television Advertising', *Genealogy* 1:4 (2017).

43 https://www.ancestry.com/corporate/newsroom/press-releases/ancestry-surpasses-5-million-people-dna-database-giving-customers-even-more [accessed 6 May 2022].

44 https://www.ancestry.co.uk/dna/ [accessed 23 February 2021].

45 https://www.myheritage.com/ [accessed 23 February 2021].

46 Eunjung Han et al, 'Clustering of 770,000 genomes reveals post-colonial population structure of North America', *Nature Communication* 8:14238 (2017), 1–12 (9).

47 See Henri-Corto Stoeklé, Marie-France Mamzer Bruneel, Guillaume Vogt and Christian Hervé, '23andMe: A new two-sided data-banking model', *BMC Medical Ethics* 17:19 (2016), doi: 10.1186/s12910-016-0101-9.

48 See Paula Saukko, 'Shifting Metaphors in Direct-To-Consumer Genetic Testing: From Genes as Information to Genes as Big Data', *New Genetics and Society*, 36:3 (2017), 296–313 and Richard Tutton (2004) '"They want to know where they came from": population genetics, identity, and family genealogy', *New Genetics and Society*, 23:1, 105–120.

49 Jerome de Groot, 'Ancestry.com and the Evolving Nature of Historical Information Companies', *The Public Historian*, 42:1 (2020), 8-28.

50 Four films at https://www.tiktok.com/@meezersqueezer [accessed 23 February 2020].

51 https://vm.tiktok.com/ZMeYrBk3M/ [accessed 23 February 2021].

52 Paige Holland, 'My dad gifted the entire family a DNA test and it uncovered a dark secret', *The Mirror*, 19 Jan 2021, https://www.mirror.co.uk/news/weird-news/my-dad-gifted-entire-family-23345500 [accessed 23 February 2021].

53 Jose van Dijck, *The Culture of Connectivity: A Critical History of Social Media* (Oxford: Oxford University Press, 2013).

54 Diana Zulli and David James Zulli, 'Extending the Internet meme: Conceptualizing technological mimesis and imitation publics on the TikTok platform', *New Media & Society*, published online 2020, https://doi.org/10.1177/1461444820983603, p. 5, 6.

55 Harmonie Ponder, 'Cat McDonald Channels TikTok Fame to a greater good', DNA Angels blog, January 23 2021, https://www.dnangels.org/cat-mcdonald-channels-tik-tok-fame-to-a-greater-good.html [accessed 23 February 2021].

56 *History Within* (U Chicago P, 2016), p. 20.

57 Critical-creative tools are being developed, see for instance Christine Scodari, 'Recuperating Ethnic Identity through Critical Genealogy', *Journal of Multidisciplinary Approach* 8:1 (2016), 47–62.

58 Jerome de Groot and Matthew Stallard, 'DNA and Family History in Australia', in 'Family History and Historians in Australia and New Zealand' ed. Sophie Scott-Brown and Malcolm Allbrook (London and New York: Routledge, 2020).

59 'Things are coming out', p. 278.

60 'Things are coming out', p. 2.

61 See Daryl Leroux, *Distorted Descent: White Claims to Indigenous Identity* (Winnipeg, MB: University of Manitoba Press, 2019) and Elizabeth Watt and Emma Kowal, 'What's at stake? Determining indigeneity in the era of DIY DNA', *New Genetics and Society* 38:2 (2019), 142–64.

62 Catherine Bliss, 'The Marketization of Identity Politics', Sociology 47:5 (2013), 1011–1025.

63 Janet K. Shim, Sonia Rab Alam and Bradley E. Aouizerat, 'Knowing something versus feeling different: the effect and non-effects of genetic ancestry on racial identity', *New Genetics and Society* 37:1 (2018), 44–66.

64 See Anne Harris, Sally Wyatt and Susan E. Kelly, 'A market in the making: the past, present and future of direct–to–consumer genomics', *New Genetics and Society*, 36:3 (2017): 197–208.

65 Tanya Evans, 'Secrets and Lies: the Radical Potential of Family History', *History Workshop Journal* 71: 1 (2011), 49-73, and Ashley Barnwell, 'Locating an Intergenerational Self in Postcolonial Family Histories', *Life Writing* 14:4 (2017), 485–93.

66 'Things are coming out', p. 285.

67 See for instance Pramod K. Nayar, "Autobiogenography: Genomes and Life Writing," *a/b: Auto/Biography Studies* 31:3 (2016), 509–25.

68 See the bioethical discussions in Ann-Marie Kramer, "The genomic imaginary: Genealogical heritage and the shaping of bioconvergent identities,", *Media Tropes* 5:1 (2015), 80–104 and Petra Nordqvist, "Genetic thinking and everyday living: On family practices and family imaginaries," *The Sociological Review* 5:4 (2017), 865–1.

69 'Things are coming out', p. 287.

70 See Amy Holdsworth, "*Who Do You Think You Are?*: Family history and memory on British television," in Erin Bell and Ann Gray, eds., *Televising History* (Basingstoke: Palgrave Macmillan, 2010), pp. 234–47.

71 Silvio Waisbord, 'McTV: Understanding the global popularity of television formats', *Television and New Media* 5:4 (2004), 359–83 and Albert Moran and Karina Aveyard, 'The place of television programme formats', *Journal of Media & Cultural Studies* 28:1 (2014), 18–27.

72 See the discussion in de Groot, *Consuming History*, pp. 194–202.

73 Claire Lynch, '*Who Do You Think You Are?* Intimate Pasts made Public', *Biography* 34:1 (2011), 108–18.

74 *Long Lost Family* 10:1, ITV, Monday 18 Jan 2021, 9 pm.

75 Lynch, '*Who Do You Think You Are?*', p. 110.

76 Debbie Kennett, Adrian Timpson, David J. Balding and Mark G. Thomas, 'The rise and fall of BritainsDNA: A tale of misleading claims, media manipulation and threats to academic freedom', *Genealogy* 2:4, 47 (2018), https://doi.org/10.3390/genealogy2040047.

77 Carol Midgley, 'DNA Journey review'. *The Times,* 11 March 2021, https://www.thetimes.co.uk/article/dna-journey-review-impossible-to-dislike-or-find-very-interesting-nmpmcf85x; Barbara Speed, 'DNA Journey, ITV review', *iNews*, 10 March 2021, https://inews.co.uk/culture/television/dna-journey-itv-review-freddie-flintoff-jamie-redknapp-907854 [both accessed 12 March 2021].

78 Speed, 'DNA Journey, ITV review', Chris Bennion, 'DNA Journey, episode 1 review', *The Telegraph*, 10 March 2021, https://www.telegraph.co.uk/tv/2021/03/10/dna-journey-episode-1-reviewno-tears-just-banter-itvs-take-celebrity/ [accessed 12 March 2021].

79 *DNA Journey* 2:1, ITV, 10 March 2021, 9 pm.

80 *DNA Journey* 2:1, ITV, 10 March 2021, 9 pm.

81 @Brad_Argent, Twitter, 10March 2021, 10:16, https://twitter.com/Brad_Argent/status/ 1369774062206083072 [accessed 16 March 2021].

82 *DNA Family Secrets*, Episode 1, BBC2 2 March 2021, 21:00.

83 *DNA Family Secrets*, Episode 2, BBC2 9 March 2021, 21:00.

84 See David Kirby, *Lab Coats in Hollywood* (Boston, MA: MIT Press, 2013).

85 Rachael Sigee, 'DNA Family Secrets review', *iNews*, 2 March 2021, https://inews. co.uk/culture/television/dna-family-secrets-bbc2-review-stacey-dooley-894811 [accessed 8 March 2021].

86 Sara Wallis, 'Stacey Dooley's DNA family secrets is like who do you think you are? on speed', *The Mirror*, 6 March 2021, https://www.mirror.co.uk/tv/tv-reviews/stacey-dooleys-dna-family-secrets-23620252 [accessed 8 March 2021].

87 See the discussion in *Consuming History*, pp. 197–8.

88 See Annette Hill, *Restyling Factual TV: Audiences and News, Documentary and Reality Genres* (London and New York: Routledge, 2007).

89 See *Consuming History*, pp. 72–3.

90 Chuck Stone, 'ROOTS: An electronic orgy in white guilt', *The Black Scholar*, 8:7 (1977), 39–41.

91 Kenney O. Ongaga, 'Henry Louis Gates, Jr.: Prolific writer and proponent of African-American literature' in *A Critical Pedagogy of Resistance* ed. James D. Kirylo (Rotterdam: SensePublishers, 2013), pp. 53–6 (p. 55).

92 *African American Lives*, episode 1.

93 *African American Lives*, episode 1.

94 *African American Lives*, episode 1.

95 *African American Lives*, episode 1, PBS, 1 February 2006.

96 *Finding your Roots*, 6:3, PBS, 7 January 2020.

97 *African American Lives*, episode 1.

98 *Finding Your Roots* 4:3 PBS, 17 October 2017.

99 *African American Lives 2*, episode 4, PBS, 13 February 2008.

100 Christine Scodari, 'Roots, representation, and resistance? Family history media & culture through a critical lens', *The Journal of American Culture* 36:3 (2013), 206–20.

EPILOGUE: FUTURE?

There are 6000 tweets sent a second. In the time you have read this sentence, 42000 tweets will have been sent. At an average, tweet length (68 characters) means 2856000 characters. The numbers are 500 million per day, 200 billion yearly. Worldwidewebsize daily estimates the size of the Internet. On the day of writing this text, it was 4.59 billion pages and a billion websites. This is the 'indexed' Internet, not including the 'dark web' or private databases. The size of the web is measured in two ways. The first is 'content' (storage capacity estimated in 2014 as 10^{24} bytes, 1 million exabytes).[1] The second is 'traffic', measured in zettabytes. In 2021, global traffic passed one zettabyte, the content of 250 billion DVDs.[2] More conventionally, the United Kingdom published 184000 books in 2013. This is globally the largest number per inhabitant.[3] Add to this the increasing ways of measuring the human being in terms of data – DNA sequencing, online family trees, genetic coding, bank accounts, online information of all kinds – or the amount of scientific data being produced and read around the world, and the amount of information is staggering. Even the amount of storage most people need for photos and documents has grown hugely in the past few years.

As a species, we are producing information at a massive rate. The 'reading' of the mass of data has led to new predictive models for social interaction.[4] Businesses and governments are scrambling as human beings seem interpretable, manageable, and – possibly – controllable through the comprehension and manipulation of information. What this boom in information has recently prompted is a discussion into just *how* this stuff might be stored. At present, we have physical libraries, and physical archives, and bookshelves. The Internet itself is 'stored' on harddisk servers around the world, using enormous amounts of power to keep them cool. Online infrastructure is expensive and vulnerable; its longevity is also limited.

The future of information storage is a crucial issue for anyone interested in the way that societies remember. A good example is family history, where public archives like census records and tax information are served online (see Chapter 6). Millions of users around the world use subscription sites such as Ancestry or Findmypast to access this public information and to create their family trees using online software. This proliferating of information raises ethical issues about access (public records being used by private companies to make a profit) and about how this data is stored, managed, and used. We all have a stake in the way that libraries and archives might work in the future, how they might be configured, how they might work, and what might be stored (and why). Do we really need to store every tweet ever sent? Making any kind of choice over what to store – what to collect, commemorate, and archive – provokes a complex discussion. Technologies for accessing – 'reading' – information need to be somehow futureproofed otherwise we will end up with huge amounts of information that cannot be used.

There are wide-ranging discussions at present, from *what* information to store (including various biobanks) to *how* to store information to *where* to store it (the Arctic; various locations in space; under water).[5] Most of these discussions are occurring within scientific communities; some technological companies are also involved. Those who have spent years thinking about memory, commemoration, and archiving – historians, archivists, and librarians – are on the fringes of the discussion. Different organisations are exploring physical ways of storing humanity's information. Physical storage on nickel disks (read by microscope) or laser-written barcodes on silica glass have been suggested. Highly experimental – and at present energy-hungry – nanotechnology looks to write information at the near-molecular level – although the use of the word 'write' is very much out of date here.[6] Nanotechnological storage would be 'read' through sophisticated microscopy and is sometimes the 'effect' of chemical change or quite complicated processes (such as Rare-earth-doped nanocrystals converting radiation (infra-red) into something 'visible'). Some of the more baroque storage models range from a flash data memory vault on the moon to private companies sending digital content to Mars to satellites orbiting the earth.

Research teams have begun to explore the possibility of using DNA to store information, called Nuclear Acid Memory (NAM).[7] It is a complex process but 'DNA-based storage might already be economically viable for long-horizon archives with a low expectation of extensive access, such as government and historical records' (p. 78). This is a way of dealing with the 'vast amounts' of information that will be generated in the future and 'far eclipse today's data flows'.[8] NAM consists of the data being 'translated' into the letters GATC, the base nucleic acids of DNA. DNA strands would then be created, which could be translated back into the 'original' by being sequenced. Researchers recently stored archival-quality versions of music by Miles Davis and have also encoded a short section of film into bacteria DNA.[9] DNA is durable and increasingly easy to produce and read; it will keep for thousands of years in the right storage conditions. This technique would enable data to survive for several million years stored as genetic material. The experimental poet Christian Bök is already teaching DNA to 'write' poetry; the band OK Go are attempting to release an album encoded on

DNA.[10] What are the implications for engineering biological matter for aesthetic or archival purposes?

There are evidently ethical, intellectual, and logistical concerns regarding these plans, the consideration of which brings us back to the central question of this book: how is genetic science changing the way that we remember and how we imagine our relationship to the past? Much of this technology is in its infancy, but developments in nanotechnology and DNA sequencing suggest that we will be seeing the results of experimentation and development applied within years. Wider questions arise about the ethics of collection and to what extent these processes will become mainstream. Print, and to a certain extent digital, have become common and reasonably democratic ways of transmitting and storing information. It remains to be seen whether future storage and writing will be as easy to access, and who will be in control of humanity's information and memory in the coming decades and centuries.

DNA is not done with history, yet, it would seem. Increasingly, the human will be known through a genetic lens, whether that be in terms of physiology or archival memory; accordingly, the way that the past is understood will transform and change.

Notes

1 Stephanie Pappas, 'How big is the internet, really?', *Live Science* (18 March 2016), https://www.livescience.com/54094-how-big-is-the-internet.html [accessed 2 December 2021].

2 Cisco Annual Internet Report, 9 March 2020, https://www.cisco.com/c/en/us/solutions/executive-perspectives/annual-internet-report/index.html [accessed 2 December 2021].

3 Alison Flood, 'UK publishes more books per capita than any other country, report shows', *The Guardian* (22 October 2014), https://www.theguardian.com/books/2014/oct/22/uk-publishes-more-books-per-capita-million-report [accessed 2 December 2021].

4 Viktor Mayer-Shonberger and Kenneth Cukier, *Big Data* (London: John Murray, 2013).

5 Melissa Guzman, Andreas M. Hein, and Chris Welch, 'Extremely long-duration storage concepts for space', *Acta Astronautica* 130 (2017), 128-37.

6 Min Gu, Qiming Zhang, and Simone Lamon, 'Nanomaterials for optical data storage', *Nature Reviews Materials* 1: 16070 (2016), https://doi.org/10.1038/natrevmats.2016.70.

7 Nick Goldman et al., 'Towards practical, high-capacity, low-maintenance information storage in synthesized DNA', *Nature* 494 (2013), 77-80.

8 Victor Zhirnov, Reza M. Zadegan, Gurtej S. Sandhu, George M. Church, and William L. Hughes, 'Nucleic acid memory', *Nature Materials* 15 (2016), 366-70 (366).

9 Evan Minsker, 'Miles Davis' "Tutu" is one of the first songs to be encoded in DNA', *Pitchfork*, 1 (October 2017), https://pitchfork.com/news/miles-davis-tutu-is-one-of-the-first-songs-to-be-encoded-in-dna/; 'Movie encoded into the DNA of bacteria', BBC News, 13 July 2017, https://www.bbc.co.uk/news/av/science-environment-40585302 [accessed 2 December 2021].

10 Christian Bök, *The Xenotext: Book 1* (Toronto, ON: Coach House Books, 2015).

INDEX

Page numbers followed by "n" indicate notes.